W9-AUN-075

SEEDTIME

On the History, Husbandry, Politics,
and Promise of Seeds

SCOTT CHASKEY

RODALE.

© 2014 by Scott Chaskey

Rodale books may be purchased for business or promotional use or for special sales. For information, please write to:
Special Markets Department, Rodale Inc., 733 Third Avenue, New York, NY 10017

Printed in the United States of America

Rodale Inc. makes every effort to use acid-free ∞, recycled paper ♻.

Permissons for excerpts used in the book can be found on page 208.

Illustrations by Liam Chaskey

Book design by Kara Plikaitis

Library of Congress Cataloging-in-Publication Data is on file with the publisher.

ISBN 978–1–60961–503–1 hardcover

Distributed to the trade by Macmillan

2   4   6   8   10   9   7   5   3   1   hardcover

We inspire and enable people to improve their lives and the world around them.
rodalebooks.com

Fair seedtime had my soul . . .

—from *The Prelude*, by William Wordsworth

CONTENTS

ILLUSTRATIONS BY
LIAM CHASKEY

PROLOGUE

Variations on a Theme

First Variation

A community that has chosen to share in fieldwork and the ritual of harvesting is entitled also to share some honest discourse. A friend, after reading one of my letters to members of our community farm, wrote me a sprightly, scolding note: "Why are you wasting your time out in those fields? You are meant to be a man of letters!"

I quite like that phrase, "a man of letters"—it's a little archaic, perhaps, but suggestive of a true vocation. Without intending to, at least consciously, I seem to have forged an identity that takes equal sustenance from the soil and from literature—plants and words spring into being like the musical notes of a red-winged blackbird returning to voice the spirit of March. My impulse to write, and to care for plants, is born out of the same impulse in the blackbird's throat—to bring the thing into being, whether it be a call or response, a comment, a question, or a celebration. I am not alone in my homeland— the intersection of language and landscape has long been a source of inspiration in American letters.

On the land and in my study, I treasure my daily encounters with both seeds and words. I originally conceived of this book as one man's experience of the practice of two disciplines, though as I began the journey, I came to realize that such an approach could miss the urgency of the story. This story is rapidly evolving and the stakes, for those concerned about biodiversity and ecological integrity, are high. I approach the story of seeds not from a library or a laboratory but with my feet accustomed to contact with the soil. As a writer, I am influenced by the words of an Aztec poet: "My heart is sprouting flowers in the middle of the night. . . . " As a farmer, my labor is inspired by the action of a plant coming to flower and producing seed for the next generation.

This book, while drawing on scholarship and research, is primarily intended to be a discovery, though I didn't foresee that our alienation from the natural world would play so central a role. In *The Bow and the Lyre*, the Mexican writer Octavio Paz observes: "The poet does not choose his words . . . when a poet finds his word he recognizes it: it was already in him." Seeds, like words, are encoded with a hidden genetic landscape that is individual to the species, and encyclopedic.

If we look at the world of plants as a topographical map, seeds are everywhere—at the foundation of mountains, at the source of rivers, and scattered over sandbanks and fields. Poetry can be found there, too—it appears within the cold water of a mountain stream, on the surface of a tidal pool, or on the face of a bank of soil dotted with the nests of swallows. As I experience almost daily, to enter the realm of seeds is to witness a kind of magical reality, one that germinates out of the soil. Much of the institutional framework that we have created to steward that soil—from the original promise of the land grant universities to the industrial agricultural system itself—threatens to sacrifice not only that magical reality but also the source of our daily bread.

In the words of Cary Fowler, head of the Global Crop Diversity Trust: "We are in the midst of a mass extinction event in agriculture, at precisely a moment in history when diversity for further adaptation is most needed." As we face the challenges of climate change and the loss of prime agricultural soils, we need a diverse seed supply to counter the unpredictable and the unknown. Instead, we continue to lose plant species—and the seeds of the future—at an alarming rate.

We can all begin to access this story through the language of the spring wind and summer rains, through the life force in a handful of silt loam, and through the tireless and careful work of some extraordinary individuals, true keepers of the earth. We should also gain some familiarity with the language and laws recorded by our universities, courts, and corporate players that either encourage or inhibit ecological actions. Octavio Paz comments that "words behave like capricious and autonomous beings." In my experience, his lyrical observation also applies to seeds.

Second Variation

For years, I have compiled notes on the mystery and magic of seeds, and my notebooks are filled not only with quotations from other writers but also with practical observations intended to enhance my abilities as a sower. That word—*sower*—may for some suggest the painter Jean-François Millet's pastoral image of a contented laborer, cloth bag draped over a shoulder, scattering seeds of grain across a tilled field. We can identify the time and place that the painter portrays—mid-19th-century France—but the image of the laborer traversing the field is timeless. As a species, we have been collecting, saving, and sowing seeds for thousands of years; the stories, the art, and the data depicting our relationship with seeds are abundant, and, like seeds, scattered around the globe. And yet, it is probable that the routine task performed in Millet's painting is unrecognizable to many 21st-century citizens.

Though our lives still depend upon the viability of the crops we choose to preserve and disseminate, knowledge of seeds has become increasingly specialized among a handful of agrarians, scientists, and activists. In an age obsessed with industrial efficiency, it is prudent to acknowledge the danger, along with the gifts, that accompanies technological advance.

The poet William Blake, writing near the dawn of the Industrial Revolution, wrote this simple, instructive proverb: "In seedtime learn, in harvest teach, in winter enjoy." The quiet wisdom of Blake's words (somewhat muffled in the year 2013) is validated by 10,000 years of humans practicing agriculture.

At one period of my agrarian education—unconsciously inspired by *The Sower*—I donned a cloth satchel to scatter a mix of grains and legumes as a cover crop over our fields. The experiment was brief; it is much more efficient to strap on an EarthWay seeder—a sort of lightweight seed bucket that is held against the chest like an infant carrier—and to turn the crank to spin out the seed. Even the tiniest of seeds, like clover, can be spun out with some precision in all directions. Throughout this task, the sower is in contact with the earth, as seeds fly out from the hopper to flick against hand, arm, and sandals, then to settle in furrows in the soil.

As the number of cultivated fields on a farm tends to multiply, the method for efficiently spreading field crops inevitably must change. At Quail Hill Farm, our community farm in Amagansett, New York, we accomplish the seasonal task of sowing cover crops by emptying bushels of oats, rye, or winter peas into the 6-foot-wide hopper of a landscape seeder, or into the 12-foot-wide grain drill. Each implement is lifted for transport, then lowered in the field and pulled behind a tractor. Still, despite the mechanical improvements, the most satisfying aspect of the job is tactile—extending your arms into the deep hopper to sift together oats, beans, buckwheat, or clover.

Similarly, when I seed the row crops by hand—acres of them—there is always a moment of fresh recognition and connection when I pour carrot seeds, or parsnips, arugula, or raab, into the well of the hand-pushed seeder. Annoyed at any number of obstacles to a smooth routine—the seeder must be oiled, the soil preparation is inadequate, a shower suddenly appears, racing from the ocean—I am delivered to the present by Blake's refrain: "In seedtime learn . . . "

❦

I HAD THE LUCK to be in London in early spring 2011 when a substantial portion of the Tate Modern turbine hall was the setting for Chinese artist Ai Weiwei's graphic exhibition *Sunflower Seeds*. The Tate Modern, on first thought an unlikely setting for a museum, is in fact the perfect canvas for Weiwei's work. This massive building, built along the Thames in two phases between 1947 and 1963 to house electricity generators, was originally known as the Bankside Power Station. The building was designed by Sir Giles Gilbert Scott, the man who also designed Waterloo Bridge, university libraries in Oxford and Cambridge, and the red telephone box. The power station was retired in 1981, and in 1995 the architects Jacques Herzog and Pierre de Meuron were chosen to convert a generating plant into gallery space for the Tate Modern's collection.

The turbine hall serves as a dramatic, enigmatic entrance into the world of modern art. Weiwei planned his exhibition of seeds to be interactive; early on, visitors were invited to walk, sit, or sprawl on top of the 100 million porcelain seeds that covered the floor of the turbine hall, roughly the size of a football pitch. It was the artist's desire that the public be allowed to touch and reflect on the productions of people and of nature, the individual and the collective, industry, art, and famine. But by mid-October of 2010, the Tate, concerned with the health implications of porcelain dust, closed the exhibition to foot traffic. Still, seen from the floor above or walking beside a football field of seeds, Weiwei's moving statement required the viewer to contemplate the paradox.

These sunflower seeds—yes, 100 million of them—were produced in the town of Jingdezhen, 1,000 kilometers from Beijing, by 1,600 Chinese artisans. Weiwei's commission brought employment to a village without work, a village that once supplied the emperors of China with objects of fine porcelain. Ultimately, his art, from conception through construction, is a commentary on the social and political landscape of his country, where industrial progress dwarfs individual rights. Actual living seeds—collected, saved, and transported around the globe—possess similar symbolic power, often as a result of and in reaction to political oppression and hegemony. Though Weiwei's sunflower seeds will never germinate in soil, each porcelain husk embodies the transformative power of art.

I paced it off—100 million sunflower seeds spread 6 to 12 inches deep occupy floor space roughly 200 feet long by 70 feet wide. Each porcelain seed was hand-painted in

Jingdezhen with a thin brush, two strokes per seed, as nature paints the seed husks of helianthus, the sunflower. Weiwei's statement is not subtle (imagine the weight, the transport of millions of clay seeds to the Tate Modern); I am reminded of another line by London's own William Blake: "Eternity is in love with the productions of time."

Looking down on a floor filled with seeds in the heart of London, I questioned what the writer Barry Lopez refers to as "our distance, both real and imagined, from the natural world." The thread that connects us is delicate. Now I travel back in my mind to my encounters with helianthus—to the sound of a seed plate revolving, to the husk that clings to the first germination, the stem that rises 12 feet out of silt loam, the brilliance of a sunflower face turning to face the sun. Weiwei's seeds, evocative of the oppressive power of the state, also speak of the promise of human hands, able to fire and paint a husk of clay or to place a seed into fertile soil.

For seeds hold promise. Imagine the weight of stalks rising out of 100 million seeds, of the pollen becoming available to bees, of the rich oil held in the flesh of the seed kernel. There is a *fabric of relationships* available to us encoded in seeds, a timeless refrain that the poet Rumi hears as "a greeting from the secret ones inside." Enfolded into this book, after years of learning to read the field, is my desire to practice listening. If you are as rooted as a 'Mongolian Giant' sunflower, or as a bending alder in a meadow, or as a bristlecone pine in mountain gravel, Rumi reminds us, "the branches of your intelligence form new leaves in the wind of this listening."

Third Variation

Seeds are often given wings, to assist their journey and to ensure another generation. It took several hundred million years or more to move from the birth of a single cell to the evolution of seed cones to the formation of seeds, and yet we now take for granted this magnificent dispersal—the language of transport of the plant kingdom—as if it were simply a mechanical inevitability rather than a mysterious gift of time. When the wings of a maple seed take off on the wind, or the cotton wings of a milkweed seed break from the pod to fly for another patch of soil, it is as if a word goes forth, an original phrase but also an echo of an ancient earthly melody.

The life force within a seed is equal to the force within a word poised for articulation; it draws the poet inside. Just as the memory held within each seed is activated and responds to light and the changing temperature of the seasons, when I write, I can feel the first breath of a melodic line before it is nourished into expression. The germination of a seed is speech, given substance.

One of the smallest seeds we have in our farm larder is also one of the most useful and graceful: clover. Of this vibrant legume, the Buddhist teacher Robert Aitken says: "Clover does not think about responsibility . . . clover simply puts down roots and puts up leaves and flowers . . . it comes forth and its response to circumstances is to give nourishment."

Between two patches of garlic (35,000 cloves in all) last autumn, I seeded a mixture of buckwheat, a little rye, and clover that would germinate in the shade of the other grains. The buckwheat took the lead and rose to 3 feet, flowering white and in abundance to attract our bees and other local pollinators. Then the rye grass stood up in the cool nights it prefers. Unseen, underneath, the tiny clover seeds put out a few leaves and began to spread across the soil. A seed contains an embryo, a miniature plant awaiting the moment of transition. Seed leaves store food within the endosperm—the seed coat—that will nourish the seedling plant when it emerges. As defined in *Flora*, an encyclopedia of plants, "A seed, in essence, is a tiny plant in suspended animation sealed up in an environment-proof coat." The beauty of clover is the slow, steady growth that carries the nourishment forward to feed the soil in another season. In my experiment, the buckwheat died off with the first frost, the rye held on over winter, and a sea of clover arrived in late spring. When I mowed the green sea in May, clover rose in measured waves to claim the soil.

❦

SEVERAL YEARS AGO, I traveled to Sand County, Wisconsin, to address a convention of land conservationists in Madison. The writing of the mid-20th-century conservationist Aldo Leopold was inspired and refined by the spare landscape of Sand County, where he came to write his "Land Ethic" not far from the meandering banks of the Wisconsin River. In this Nordic riverine place, Leopold wrote that land is not merely soil, "it is a fountain of energy flowing through a circuit of soils, plants, and animals."

To offer a blessing for the first organic meal served to this annual "rally" of 2,000 conservationists, I turned to the language of Leopold and of the poetess of Wisconsin's Black Hawk Island, Lorine Niedecker. Quiet and subtle, their words rise and fall with the breath of the land and water they were part of. When Niedecker writes "Sun, turn the earth once more . . . ," she is aware, as is Leopold, that "the circuit is not closed . . . it is a sustained circuit." Hold a seed between finger and thumb and you are part of the circuit that leads from the soil to the food to the table. After a visit to Aldo's shack (not yet modernized, luckily), I sat under a stand of trees to write of this resonant place, and a shower of pine needles came down on the wind to cover my words. The harmony created by cranes in the distance and a small leather-bound notebook laced with pine quills became a basis for this book.

Now I am also compelled to write in response to the urgent needs of planet Earth and to recover some sense of balance in our understanding and care for the land, in how we choose to manage and distribute seeds and to grow the food that sustains us. Though I am a sower, I also enter the world of seeds through the voices of other writers, "fostered alike by beauty and by fear" (Wordsworth's phrase). "A strong song tows us," my former teacher, the poet Basil Bunting, wrote, and "blind we follow, rainslant, sprayflick, to fields we do not know."

As we travel farther into space and crowd the earthly space we have, this is also the time to recognize the beauty in the architecture of our soils and the transformative seed language born of these soils. If we accept our role as "mere citizens" of the biological community, as Leopold defined it, then the genetic wisdom carried in a seed kernel is an invitation to nurture our collective future, not to manipulate ecologies or markets. Miguel Altieri, the noted agroecologist, has written: "The kinds of agriculture with the best chance to endure are those that deviate least from the diversity of the natural plant communities within which they exist."

I listen to the call of a mother wren in beech woods, the crack of a hickory branch in a thunderstorm, the race of a shard of ice from a frozen shelf of sand, and these sounds speak to me of what we have been given and who we are. As a farmer, I am a reader of the natural world, and my response is to listen with a "passionate intensity" (W. B. Yeats) and to give back nourishment. As a writer, I recall Samuel Johnson's beautiful phrase "Words are the daughters of earth," and I search for fair seedtime to find word roots that will blossom and reveal some greenery.

Planting words, like seeds, under rock and fallen logs—
letting language take root, once again, in the earthen silence of
shadow and bone and leaf.

—David Abram, *The Spell of the Sensuous*

CHAPTER 1

Furnished with Wings and Plumes

ENCAPSULATED IN EACH SEED is a story, a story held in a state of rest until released. Only with significant patience and effort can we interpret this language, which gradually is revealed as the cotyledons, or first leaves, unfold from a seed's invisible center. A plant's coming into being, or maturation, is such a quiet progression that we tend instead to focus on the fruit, the colorful prize of production and the vessel of taste. To grasp the whole story, however, we will have to look at the structure of a flower,

1

how plants have evolved to attract pollinators, and how a flowering plant produces seed. Our entire food supply is a gift of the angiosperm revolution—the magnificent event that introduced flowering plants to the world 140 million years ago—and our health and food futures are entwined with the way in which we choose to nurture or manipulate the seeds of that natural revolution.

I agree with the sentiment expressed in the opening pages of Henry Hobhouse's 1985 book *Seeds of Change*: " . . . the world cannot evolve solely through actions consciously willed by man." Hobhouse urges us not to dismiss plants as "less than fundamental to history." More recently, Michael Pollan, in *The Botany of Desire*, suggests that the plant world may have far greater control over us than we could ever imagine. Most of us, when we do think of this relationship at all, presume that we comfortably inhabit the role of conqueror.

Years of working in the field with plants and planting seeds each season have taught me otherwise. Often, when I enter a field to inspect or to cultivate a given plant species, I am "called" to tend to another species I had forgotten or had passed over. Garlic is a crop that seems to dominate my entire work schedule—at least until I have weeded, fertilized, and mulched to the satisfaction of this allium (*A. sativum* var. *ophioscorodon*).

It is because of my daily work in the fields that I am aware of this other time: seedtime. A farmer relies on both intuition and experience to build fertility into a field, to bring a crop to fruition, or to establish a field of grains—oats, buckwheat, rye, or red clover. It is much more difficult to discern the precise syllables latent within a seed embryo. To do that, or at least to attempt to, I am led beyond the crop rows to the adjacent hedge-row, where I can observe the beauty and diversity of wild seeds.

We need wildness, but wildness needs space. It also, increasingly, needs us. In their book *Shattering*, longtime agricultural activists Cary Fowler and Pat Roy Mooney praise the value of diversity within agricultural systems:

> The future of agriculture and the very survival of crops depend not so much on the fancy hybrids we see in the fields, but on the wild species growing along the fence rows, and the primitive types tended by the world's peasant farmers in the centers of diversity.[1]

Jack Harlan, agronomist, plant explorer, and author of *Crops and Man*, is even more vehement about the role of uncultivated species in our survival: "Wild relatives stand between man and starvation." His point is that plant breeders, professional and amateur alike, will always require genetic variation when selecting and breeding plants to build resistance to pests and disease, and that variation is often to be found in species growing on the edges of cultivated fields or on undeveloped or preserved land. The future of our food is quite literally in the hands of man, to conserve wildness not only in land but also in

---

1   A hybrid, in this specific sense, is the progeny of a cross between two varieties or races of the same species, which themselves have been produced by repeated self-fertilization or inbreeding. Hybrid plants and the centers of diversity will be discussed in a later chapter.

plants. These wild relatives provide resilience and inspiration to our inbred cultivars (those plants developed through systematic breeding).

The story of seeds is now so interwoven with our own story, and of how we have chosen to manipulate our environment, that it behooves us to trace the evolutionary history of flowering plants and seeds through all the variations of dispersal. In my imagination, I am swept up on the wind within a galaxy of pollen grains—though in my study, I am content to follow Henry D. Thoreau through his work "The Dispersion of Seeds," as he records how a seed journeys by the "agency of wind, water, and animals." (*Faith in a Seed*, Thoreau's neglected manuscript published some 125 years posthumously in 1993, includes this essay.) The opening chapters of Darwin's *On the Origin of Species* (published in 1859), a book that strongly influenced Thoreau, also discuss the dispersion of plants and animals across the earth.[2] Darwin's description of this process reveals breathtaking observational precision:

> Seeds are disseminated by their minuteness—by their capsule being converted into a light balloon-like envelope—by being embedded in pulp or flesh,

2   As I was reviewing this chapter, I traveled once more to London and met a family friend in Piccadilly. She revealed that her literary agency was located in a nearby building still occupied by the oldest publishing house in England, John Murray. Sensing my interest, she invited us to meet the ninth-generation descendant of Mr. Murray. In a spacious room packed with leather-bound tomes, he introduced us to a long line of authors and exclaimed, while pointing to an austere oil portrait above the central fireplace: "Here is perhaps our most well-known of writers . . . the author of *On the Origin of Species!*"

formed of the most diverse parts, and rendered nutritious, as well as conspicuously colored, so as to attract
and be devoured by birds—by having hooks and
grapnels of many kinds and serrated awns, so as to
adhere to the fur of quadrupeds—and by being fur-
nished with wings and plumes, as different in shape
as they are elegant in structure, so as to be wafted by
every breeze.

The variation in the shape and structure of seeds is a fascinating story in itself, and the diversity in their size is impressive. Orchids reproduce via microscopic seeds—one tropical
orchid seed capsule can contain 3,750,000 seeds. At the other
end of the spectrum, the planet's largest seed, the Seychelles
nut (coco de mer), can weigh up to 45 pounds.

꙳

THROUGHOUT MOST OF the history of agriculture, each farmer
was by definition a seedsman. It was the man or, in many cases,
the woman in the field who selected the strongest plants and
collected and saved the seeds to ensure another harvest. The
relative success of this vocation was as unpredictable as life
itself—given the powerful influence of weather, affecting both
plant growth and seed storage. But the point is that the knowledge of plants and food production was a shared knowledge
among growers and eaters. Gradually, as we approached the era
of industrialization and specialization, this fluency in our shared

ecology slipped away, and with it an incalculable biodiversity. Seed companies eventually replaced farmers in the field as the keepers and purveyors of seed. As seed production became more centralized, on-farm (in situ) breeding and seed selection diminished, and the indigenous wisdom of generations began to fade. How many of us can name the difference between an open-pollinated plant and a hybrid, let alone understand the implications of our present industrial systems for our food supply? If we retrace the story of seeds to the waters and soil of origin, as if observing the emergence of seed leaves from an embryo, we will glimpse a shared identity. We are, after all, fellow travelers on this earth and dependent on each other.

Our public debates—concerning health care, chronic disease, and obesity—are just now beginning to acknowledge the role of food and how we grow it. A recent letter from Slow Food USA begins with this:

> Our broken food system is at the heart of nearly every crisis we face. The way we eat and the way we farm are devastating our society, our health, and our planet. Americans of all ages are sick from diet-related disease, while overwhelming debts are pushing farmers out of work and pollution from industrial farms is contaminating our water, soil and atmosphere.

The message is dire, but a "blessed unrest" (Martha Graham's words, the title of a recent book by Paul Hawken) is stirring, led by activists addressing an entrenched agricultural system and searching for a deeper ecological understanding.

The value of conserving biodiversity cannot be overstated. (Biological diversity refers to the abundance of organisms that populate our earth; this includes species, genetic diversity, and ecosystem diversity.) Biodiversity is the source of our food, of the animals and plants that we have domesticated, and of plant-based medicines. Our increasing tendency to homogenize all aspects of our ecosystems limits our ability to adapt to ever-changing conditions of climate and culture. As sophisticated as we presume ourselves to be, it is estimated that to date we have identified only about 10 percent of all species on earth.

It is incumbent upon us now to preserve those species that we have identified and to protect habitats where those undiscovered may still flourish. In February 2008, after 20 years of preparatory work, the first world seed bank was christened in Svalbard, Norway. The stated purpose of this seed bank, like that of the great Russian plant explorer Nikolai Vavilov's institute in early-20th-century St. Petersburg, is to preserve genetic diversity and protect our global food supplies. Svalbard's mission interprets the work accomplished on small farms and gardens throughout the globe and the efforts of a new breed of "bioneers." Seedsman and plant breeder Bryan Connolly has said, "Seed saving done well is at the heart of regenerating US agriculture." Bill McDorman, director of Native Seeds/ SEARCH, stresses that it is up to ordinary farmers and gardeners to save our diversity while we still can, "one farmer, one field at a time." Tom Stearns, founder of High Mowing Organic Seeds, a Vermont company whose catalog includes only organic seeds, comments: "I see my work focusing on helping people rebuild their local food systems. Seeds are an important, yet easy tool to recognize the [value] of these systems, and seeds are one of the mediums by which this message can be conveyed." Ethnobotanist and author Gary Paul Nabhan writes: "Modern agriculture has let temporary cheap petrochemicals and water substitute for the natural intelligence—the stored genetic and ecological information—in self-adjusting biological communities." A significant counterculture of seed activists is now conducting field research based on that natural intelligence.

꙳

WITNESSING THE TRANSITIONS of the seasons, I hope to keep my eyes open to what the day reveals—not as an aesthetician but as a seedsman inspired by the stories held within silt loam and the fiber of roots, stems, and leaves. As Thoreau described it: "The scholar's and the farmer's work are strictly analogous." In "The Dispersion of Seeds," he praises the Old World's recognition of the essential, inherent value of seed, contrasting it with the inability of the New World citizen (in the 19th century) to recognize that a tree is actually born of a seed. My work in the fields, and in the classroom, is dedicated to revealing this connection: the interrelation of things.

... where alack,
Shall Time's best jewel from time's chest lie hid?

—Shakespeare, Sonnet 65

CHAPTER 2

Time's Best Jewel

DIGGING THROUGH STORAGE BOXES in the closet of a back room, my wife, Megan, unearthed some seeds this winter. The envelope containing the stored energy of sunflowers bore sufficient evidence to piece together a story. Several black-and-white photographs transported us back to the first garden that we shared, circa 1979, tucked into the hills of the Southern Tier of New York State. We were there as caretakers, at the home of a fiction writer, a college teacher of mine, and I remember spading and seeding and mulching in the garden at the first light, and then again when darkness came. The photographs of dancing sunflowers—with their laughing, spiral faces—revealed the location, but the real story, and the more complex history of helianthus extending back unknown years, was embedded in the white-and-black-striped seeds. Seeds often prefer to hitch a ride on the most available passing animal; these had chosen us, and we had carried them, boxed, by Pontiac and Saab and from house to house for 30 years. Of course, I was curious—I tucked a few seeds between layers of paper towels, added some water, and then placed the encased potential in our south-facing window. . . .

꙾

IT IS DIFFICULT to imagine a world without flowers—and therefore without seeds—but that world existed, and without us to imagine it. We rarely give a thought to the role performed by seeds in our world, yet nearly everyone has heard at least one incredible story of their acrobatic nature. For we must remind ourselves—because they do not appear to be so—that seeds are actually living things. The purpose of a seed is to ensure the survival of the parent plant; as this remarkable capacity evolved in plants, Earth's landscape was transformed. The genius of evolution is revealed when a seed awakens, after one, one hundred, or more than a thousand years.

In 1940, when a bomb struck the London Museum of Natural History, a diverse collection of seeds was subjected to fire and water. The seeds' response was natural to this man-made combustion—they germinated. Collected from sites scattered around the globe, some had been stored for 150 years prior to this elemental shock.

Viable seed for a plant known as Indian shot (*Canna indica*) was found in a tomb in Argentina, embedded in a walnut shell as part of a rattle necklace; radiocarbon dating revealed the seed to be 600 years old. In 1995, seeds of the sacred lotus (*Nelumbo nucifera*), uncovered in a dry lake bed in China, were successfully germinated, and radiocarbon dating placed them at about 1,300 years old.

In 1982, in the course of excavating a 2,000-year-old settlement in Japan within a grain storage pit, archaeologists unearthed a single unusual specimen. Planted and watered, this seed revealed itself to be a magnolia. Following this remarkable germination, the plant was assumed to be an example of a common wild species. But 11 years later, when a flower magically appeared, this magnolia unfurled eight petals, unlike the characteristic six petals of *Magnolia kobus*. Roused from a 2,000-year slumber, this individual has been identified as a lost species suspended in time within the coat of a seed.

In 1967, seeds of the arctic lupine (*Lupinus arcticus*) found in the frozen tundra in Canada were successfully germinated, and these specimens—found in a lemming burrow—were originally thought to be 10,000 years old; later, radiocarbon dating revealed the seeds to be modern contaminants. More recently, in June 2005, Israeli researchers were able to germinate some seeds of the date palm (*Phoenix dactylifera*) excavated by archaeologists in the 1970s in the ruins of Masada, a fortress built by King Herod in the Judean desert. The seedling was nicknamed

Methuselah. Radiocarbon dating established that the seed was 2,000 years old.

These seed stories seem to border on fiction and naturally attract attention, but it is perhaps misleading to focus on the fantastic. The story of how seeds came into being, and of how they continue to embody life itself, is the real archival text. Our own human story is interwoven into that text; though we arrive in a later chapter, our influence is substantial.

꽃

ABOUT 2 BILLION years ago, the first plant life, a distant relative of algae referred to as "blue-greens," appeared in the waters covering the earth. For a billion years or more, this original life-form continued to reproduce by cellular division, one dividing into two. Gradually, a few intelligent organisms discovered a way to make use of sunlight as an energy source, and this increase in the ability of cellular life to produce food led to more sophisticated life-forms. Still, until relatively recent times—geologically speaking—early plant life was reliant on water to reproduce. Sperm cells in search of female eggs had but one choice, and that required the art of the backstroke or the crawl. In most plant reproduction, as is true for our own species, each parent contributes one set of chromosomes, which contain the genes that define the organism. Sexual reproduction proceeds in the same fashion everywhere—a sperm must fertilize an egg cell in order to produce offspring. When primitive plant life first reached out for a rock ledge, or a slice of stone at the edge of a body of water, about 445 million years ago, a new relationship had begun, and this gesture would eventually lead to a world populated by seed plants.

The first seed plants, known as spermatophytes, appeared 360 million years ago at the end of the Devonian period. These plants differed from the spore-bearing ferns that preceded them in two significant ways. The new seed plants held on to their female spores (megaspores) within a kind of container—the megasporangia—instead of dispersing the spores on wind or water; they also were able to cover their megasporangia with a protective layer known as an integument.

As the megasporanium evolved into an improved organ, the ovule, a more efficient means of transport for sperm was called for. Pollen (from the Latin for "fine flour") evolved to answer the call. Pollen grains are tiny spores that are produced in pollen sacs (known as the archespore). When the pollen sacs open, what we call pollination begins—male cells are released into the air or water in search of female (egg) cells. In the seed plants, a fertilized egg cell will develop within an ovule and remain at rest there for an undetermined amount of time; spore-bearing plants are incapable of this.

The warm climate of the Paleozoic era—which encompassed the geological time periods of the Devonian and Carboniferous—favored the continued dominance of giant spore-bearing plants. Now-extinct relatives of club mosses, horsetails, and tree ferns flourished in the prehistoric swamps. In the Permian (geologic) period (248 million to 290 million years ago), a shift in landmasses initiated a global cooling and a mass extinction of land and ocean creatures. Spore-bearing plants suffered, while seed plants (spermatophytes) came into their own—this was truly their time to flower. And flower they did; today they make up 97 percent of all land plants.

As seed plants continued to adapt, they became more adept at pollen transfer. Two hundred and ninety million years ago, the first "naked seeded" plants, the gymnosperms, appeared on Earth. The 20th-century American botanist Donald Culross Peattie referred to them as "plants without wombs." Gymnosperms had evolved a way to produce seeds within a cone; they are described as "naked" because the seeds are held only loosely within scales. The innovation of the gymnosperms was to refine the method by which male cells within pollen grains could fertilize female cells found (with some luck) in another cone. Once fertilized, an egg can remain within a cone until the woody shell splits to release the seed. The right visitations of sun and water and temperature lead to germination and the beginning of another generation.

During the Mesozoic era that followed, cycads—one of the early gymnosperms—multiplied to such an extent that this era is sometimes called the "age of cycads and dinosaurs" (cycads, conifers, and ginkgos were the food plants of dinosaurs). Cycads still survive—in locations ranging from Cuba to Australia—though they are limited to only 290 species. It is amazing that they are relatively unchanged over a period of 200 million years.

Conifers ("cone-bearing"), the gymnosperms we are most familiar with—including pine, spruce, and fir—also flourished in the Mesozoic, with a much greater diversity of species than now exists. Still, conifer forests presently cover 25 percent of the land surface of the earth—they are dominant in cold and dry conditions.

In the mountains of eastern California, 4,500 years after a seed fell from the scales of a bristlecone pine in a barren landscape, we can still touch the wood tissue of a rugged individual conifer at an altitude of 10,000 feet. Out of a seed of a similar size, one that can be pressed between a finger and a thumb, the world's largest organism germinates—the giant sequoia. Rising 300 feet in the air along the coast of California are trees that weigh more than 6,000 tons, alive and well 2,500 years after first sending down rootlets into soil.

The production of fantastic amounts of pollen was a requirement for the gymnosperms to multiply over the earth— one birch catkin may hold more than 5 million grains of pollen. We take it for granted that insects, especially honeybees, have always searched for and collected this exquisite dust. But the original purpose of pollen was to enclose the male sperm, and to provide a food source, until a female cell could be found. The fortune of the first seeds was intimately linked with the force of the wind.

And the wind, without forethought, carried the naked seeds of gymnosperms thousands of feet in the air and up to 3,000 miles in every direction. The biologist and founder of the Nature Institute, Craig Holdrege, has noted: "Pollen can circle the world in global winds." I like to picture the first seeds closely communicating with the wind, as in this passage from the Brihadaranyaka Upanishad: "Then the wind makes room for him, like the hole of a carriage wheel, and through it he mounts higher."

From the birth of the gymnosperms, we must travel forward millions of years until we reach the time roughly 140 million years ago when we encounter the "abominable mystery" noted by Darwin: the rise of the angiosperms (the word means "encased seeds"). Beginning with flowering plants, over a period of about 30 million years an incredible variety of seeds and fruits came into being. Earth underwent what has been called "an ecological revolution." The river of life began to flow with more velocity; with flowering plants, the story of seeds was enlivened to include pollinators and transporters—insects, mammals, and a diversity of bird life. Before the arrival of the angiosperms, various primitive mammals and birds traveled within the canopy of Cretaceous forests. But as American anthropologist Loren Eiseley points out: "They were waiting for the Age of Flowers." The angiosperms were developing a means to surpass the dominance of cycads and gymnosperms; they were increasingly able to move beyond the canopy and to populate widely varied landscapes.

It is estimated that between 250,000 and 300,000 angiosperms exist in the world today (reported figures seem to vary—seed morphologist Wolfgang Stuppy quotes the number as 422,000 species). The ethnobotanist Gary Paul Nabhan reminds us that we are in debt to the angiosperms:

In one way, a retrospect on this revolution must humble us. Few of the edible, nutritional characteristics of the seed plants that now sustain us evolved for our benefit, under selective pressure from our forebears or through conscious breeding by scientists. We are literally living off the fruits of other creatures' labors—those of the birds, bugs, and beasts that loosely coevolved within seed plants over the last hundred million years.

When the first flowering plants appeared—at the end of the age of dinosaurs—they still relied on the wind to transport their pollen. But gradually, and increasingly, things with wings and four legs came to serve as agents of transport. Creatures were lured by color, by scent, and by the promise of rich food; flowering plants entered into a coevolutionary relationship with animal and insect species.

At the center of the mystery that intrigued Darwin is the physical structure of a flower.[1] To ensure pollination, a gymnosperm situates ovules on a branch or along the edges of leaves, naked, though borne on a cone. By contrast, the angiosperms evolved a system to protect the ovule and to escort pollen to the chosen place. The ovule of a flower is contained within a carpel. Wet, receptive tissue on the carpel's surface, the stigma, provides the perfect place for pollen grains to germinate. Pollen then pushes its way into the flower's womb, where a seed can

---

1   The etymologies of certain flower parts are fascinating and instructive. *Anther* is derived from a Greek word meaning "flower"; *stamen* from the Latin for "thread"; *pistil*, Latin, "pestle"; *stigma*, Greek, "spot"; *style*, Latin, "stylus"; *ovary*, Latin, "egg"; *corolla*, Latin, "little crown"; *petal*, Greek, "thin plate"; *calyx and sepal*, both from Greek, "covering."

be formed. Whereas gymnosperms are required to generate and release vast amounts of pollen in order to reproduce, angiosperms, by inviting insects to join in the game, have perfected a more precise choreography.

The spread of grasses (also angiosperms), which would come to number more than 6,000 species, literally changed the color of much of the landmass of earth. With the profusion of cereal grains and this new supply of concentrated energy, a diversity of larger herbivores coevolved to graze the land. Mammoths, horses, bison, and more warm-blooded animals evolved in answer to the abundant food source provided by flowering plants—a remarkable interdependence that we have inherited, if we seldom recognize it. Grains are part of the grass family, *Gramineae*, and thus belong to the larger classification of flowering plants. The flowers produced by this family are very small, but the seeds (which we sometimes call berries) provide the predominant share of our daily bread—wheat, oats, barley, and rye, among others. We have refined and transformed the harvest of grains, but this family has nourished all forms of creatures for hundreds of thousands of years.

By offering up pollen, nectar, or sometimes a fruit or the nutrition encased in a seed, the angiosperms altered the balance of the earth as we know it. Darwin used the word *abominable* to express his perplexity that a family of plants had made such an imaginative evolutionary leap—they appeared quite suddenly, with no obvious predecessors, and were uncommonly adept at forming relationships. To appraise the gift of the angiosperms, I prefer to borrow a word used by Shakespeare: *beauty*:

> *Since brass, nor stone, nor earth, nor boundless sea,*
> *But sad mortality o'ersways their power,*
> *How with this rage will beauty hold a plea,*
> *Whose action is no stronger than a flower?*

This action, like the efficient mechanism within a flower that ensures another generation, may indeed be "time's best jewel."

❊

AFTER SEVERAL WEEKS of careful tending, the Southern Tier sunflower seeds I had placed in our south-facing window, asleep for 30 years, refused to open up. It was cold when I began the experiment, and perhaps I was a bit inconsistent with my watering—sunflowers, after all, wear a rather thick coat. The weather has warmed, so I have carried a few gems of this helianthus to our farm workshop. With some ceremony, I break apart several seed coats and slip the oval seeds into a few pots of our finest seed mix, made with our own aged compost. I select a choice corner of the heated greenhouse, and as the red-winged blackbirds announce spring, I lift the wand. . . .

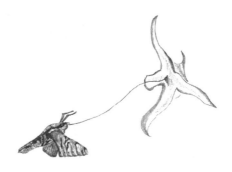

The mystery of action: We are seeds.

—from "Seeds and Rain," Rumi, translated by Coleman Barks

CHAPTER 3

"A Seed Is a Book"

WE HAVE A CUSTOM at our community farm that if three people request a particular vegetable or flower, we will seed it and introduce the plant into our fields. The farm has grown considerably in the last 24 years, so now more than 700 people have the chance to make such a request. Given that we already list more than 500 varieties, it is obvious that we are either generous or a bit rash.

Arctium lappa, or burdock, is one such crop. I got to know this tenacious root some years ago when my family was introduced to the macrobiotic diet. As a gardener/farmer, I am an earthy sort, so I felt an immediate affinity to this most yang of cultivars. The thing itself resembles a very long carrot, though the leafage is more rhubarblike; it rises out of the soil at a snail's pace as if to announce that the gardener must exercise extreme patience—the edible portion will not mature for another 120 days. The seed is unlike the smaller carrot seeds—it is more substantial and rounded, and thus it is difficult to identify which seeding plate will be the right fit.

The year when we first harvested burdock, one of the three members requesting it, a Japanese woman, brought me a delicious dish of *gobo* prepared in traditional fashion. The sturdy burdock root had been sliced very thin and sautéed with scallions, garlic, and sesame, then simmered in a covering of mirin and shoyu. Her face was radiant when she presented it to the farmer, and thus we continue our liberal policy of variety selection, hoping to stimulate an increased ecological awareness among our community members.

Tenaciousness is a quality I respect and trust. My first son, Levin, a hardworking and thoughtful man, is a Taurus (sign of the bull), whose birthday in May just precedes that of my poet-mentor, Milton Kessler. I remember Milt once exclaiming, with significant pride: "I'm the most tenacious bastard you will ever meet!"[1] The reader is free to interpret his meaning, but if you want to feel the word at work, try digging up a burdock root. An ordinary fork is useless; a serious digging spade with an extended metal blade will give you an advantage; a spading fork will improve your confidence; but to get to the end of the root, you must act—mind and body—in

---

1   Milt's tenacity fed his work; he was the most gifted of teachers, and he knew how to compose an emotive line: "Love, death is everywhere, life just right here."

sympathy with the way in which burdock has been programmed to grow. Deep, in search of nutrients.[2]

Let us go back to the seed, which appears at the end of a long stalk—at least in the cultivated species we plant—embedded in a thistlelike flower. Burdock has devised a most effective means to procreate. When the flower dries and the seed is mature, the seed head, bristling with sharp hairs, is ready to ride on anyone—clothed in fur or cotton or feathers—who may happen to pass by. So efficient is this method of dispersal that this member of the Asteraceae family was actually the inspiration for the invention of Velcro. In the 1940s, a Swiss inventor by the name of George de Mestral, returning from a walk with his dog, each covered with burrs and seeds, decided to take a closer look under a microscope. When he examined the hook-and-loop system that this plant had invented to aid in seed dispersal, he imagined a way to translate such efficiency into man's desire to fasten. Voilà, another advantage gained—for travelers everywhere—as a result of the angiosperm revolution.

⚘

DONALD CULROSS PEATTIE refers to the days of "conifer might and cycad ascendancy" that followed the Mesozoic era—before the renaissance of plant life, the rise of flowering plants.

---

2   Leo Tolstoy, upon finding a burdock shoot still struggling to grow in the middle of a recently plowed field, wrote in his journal: " . . . black from dust but still alive and red in the center . . . It makes me want to write." My first son's name, Levin, derives from a character in Tolstoy's novel *Anna Karenina* (Tolstoy himself), which further convinces me that our connection has surfaced from deep in the soil.

Angiosperms have been identified in fossils that date back 120 million years, though it is likely that flowering plants existed even earlier, as I have pointed out; by the late Cretaceous period, 65 million years ago, an increasingly diverse dance of seeds encircled the earth, until eventually more than 250,000 species of flowering plants had put down roots. But flowers could not reproduce and multiply without some creative support, and this arrived in the form of insect partners, or "mutualists."

Toward the end of their book *The Forgotten Pollinators*, Stephen Buchmann and Gary Paul Nabhan include an appendix that lists categories of pollen vectors (or pollinators) and the corresponding percentage of angiosperms pollinated. Wind, an abiotic vector, tops the list and is responsible for pollinating more than 8 percent of flowering plants (the authors note that the numbers are estimates). Moving down the list of vectors, we encounter bees, butterflies, birds, and bats; beetles are by far the busiest, pollinating 88 percent of angiosperms. In another appendix, pollinators are identified in relation to major crop plants. For instance, wind is the known pollinator of wheat, rice, maize, barley, and oats—the grains—while bees visit soybeans, sweet potatoes, chickpeas, sunflowers, cottonseed, and chiles. The wind pollinates walnuts; bees attend mustard; thrips and bees are attracted to peas; wind, bees, flies, and bats all are drawn to coconut. Wasps and flies are linked with star anise, moths with papaya, the wind with quinoa. A graceful word, *anemophily*, meaning "wind loving," describes the act of the wind pollinating flowering plants and gymnosperms.

These are the mutualists, though Buchmann and Nabhan distinguish between the "obligate mutualists"—where a fraternity exists between one pollinator and one plant species—and the majority of mutualists, which are the offspring of "diffuse coevolution." By this scenario, both pollen vector and flowering plant possess some options, an extended freedom in both space and time. When we consider this interaction of species and environment, diversity is not so much an abstract concept but rather an evolutionary expression.

⚘

IT WAS NOT UNTIL 1682 that a botanist by the name of Nehemiah Grew made the discovery that pollen must reach the stigma of a flower in order for seeds to form. The great 18th-century Swedish botanist Carl Linnaeus—who gave us a new system of botanical nomenclature—arrived at the same conclusion, though both men mistakenly believed that pollen (sperm cells) were transferred to an ovule (egg cells) only within the same plant.[3] In fact, the genius of pollen is that it can travel great distances to accomplish a single task: fertilization. Great forests were born throughout the surface of the globe because of this evolutionary efficiency. One hundred years later, the German theologian and naturalist Christian Konrad Sprengel observed that in some cases the stamens (male expression) of a flower can actually prevent pollen from reaching the flower's stigma

---

3   Carl Linnaeus inherited his surname—uncommon in Sweden in his time—from his father, a name chosen in honor of an ancient linden tree that grew on the family land.

(female expression). He seems to be the first to have recognized that insects are intimately involved in pollinating plants, though he didn't realize the significance of cross-fertilization. Around 1859, Charles Darwin pointed out the importance of pollinators in distributing pollen from one plant to another.

There is a well-known story demonstrating Darwin's intuition that is worth retelling. The publication of his work inspired some protest, of course—his theory of evolution was viewed as a threat to religious doctrine—and certain specific observations and hypotheses were met with skepticism. If his theory of adaptation was correct, one critic inveighed, what kind of insect or animal pollinator could successfully reach 11 inches into the nectary of the comet orchid, a plant native to Madagascar? Darwin predicted the existence of just such an insect, though it was not until after his death that a giant hawkmoth (*Xanthopan morganii* var. *praedicta*) was discovered—with a tongue long enough to reach down the entire distance of the orchid's nectary to the base of the floral tube!

Over millions of years, flowers have evolved to ensure their own survival among a diverse ecology. Only in the last 150 years have we learned enough of their innate intelligence to manipulate their inheritance for our own gain—previously, we simply shared that inheritance. Today, intoxicated by the feat of our technological intervention—the ability to transfer genes from one species to another—we ignore the genius and generosity of the flower/insect/pollinator coevolution.

❁

BY EARLY SUMMER, in the hedgerows and wild areas adjacent to our farm fields, the stalks of a strikingly plain yet elegant plant rise above the other vegetation. When the unusual flowers appear—globelike and pink-candy colored—I am often asked to name them by observant farm members who come to harvest in the crop fields. I know the plant; I've farmed beside it for more than 20 years. And yet when I answer, "Milkweed," the response is typically, "No, it can't be."

Almost everyone is familiar with milkweed's prominent seedpod—one that holds on to the upright stalk far into the autumn, gradually parting to reveal a plethora of winged seeds perfectly formed to ride on the wind. Thoreau claims to have counted 134 seeds in one milkweed pod and 270 in another, and he has a lovely description of the vessel itself: "a faery-like casket shape, somewhat like a canoe."[4] *Asclepias*—as the genus is known—remains a popular topic of field conversation because it serves as the primary food source for the migrating monarch butterflies—and almost everyone shares the sentiment expressed by Rachel Carson, who wrote, wistfully, "I remember the Monarchs. . . . " It is the shape-shifting transition from flower to seedpod that causes the plant's misidentification—the parts don't quite make sense.

The artist and plant lover Anne Ophelia Dowden spoke

---

4   Thoreau captures the promise at the heart of the milkweed plant when he writes: "Who could believe in prophecies of Daniel or of Miller that the world would end this summer, while one milkweed with faith matured its seeds?"

of the milkweed flower as "intricate and confusing." It is almost impossible to describe the process by which an insect pollinates a milkweed flower without the assistance of time-lapse photography. Because I cannot improve on her description, I quote the pithy lyricism of Anne Dowden describing the flower "perfectly constructed for cross-fertilization":

> Its petals and sepals are turned down, and the insect pollinizer alights on a corona, or crown, of five hoods, or nectar tubes. From the center of each hood projects a tiny horn, and the whole crown resembles the setting for a ring. As the insect grasps the crown with its legs, preparing to thrust its tongue into a hood, its feet almost invariably slip into the slits between the hoods. What happens next can be best demonstrated by pushing the point of a pin into one of these slits and pulling it gently upward. Like the leg of the departing insect, it will emerge from the top of the slit carrying a tiny pair of pollinia— waxy masses of pollen held together by slender arms that end in a black, sticky disk. If the pin is then pushed into a similar slit of another flower, and again pulled upward, the pollinia will be dragged after it and will be rubbed against the stigma inside the slit. In this manner the insect pollinates the milkweed.

The insect thus drinks the nectar of the flower and transports pollen to another. This fantastic process, which occurs over and over again throughout a vast geographical area, ensures the mutual survival of the most commonly known butterfly and of its common food source, the *Asclepias* genus. I would recommend crouching in the hedgerow in June to observe the intricate invention of centuries—the chorus of time involving sweetness, sustenance, wings, petals, pods, and seeds—available for in(tro)spection, assuming we can conserve it.

⚕

PLANTS ARE ROOTED in the earth—surely an evolved relationship we could learn from—so we tend to think of seed dispersal as a gesture disconnected from the mother plant. However, when a seed falls, or flies, or is carried away from the plant, it is an active expression of the whole plant. Part of a living plant is designed to travel: Plants can convince an insect, an animal, or a human to assist in the journey; they may possess fruits that disperse themselves; the entire plant may become mobile; or they may have evolved to be carried forward on wind or by water. The small-leafed lime tree (Linnaeus's namesake) produces a cluster of nuts that hang from a shoot (the inflorescence) that extends from a branch; a larger leaf bract serves as a wing to carry the seeds to the ground, just as the familiar seed "copters" carry the progeny of maples to the earth. When birds dine on various fruits, the seed may travel through their body to be deposited in a hedge or in the grasses under a wire. Wherever we have fences on our farm (to protect the crops from deer), we find berry canes sprouting up, and pokeweed, a handsome spreading plant that offers an abundance of juicy berries for small birds (poisonous to us). Asparagus spears rise from the soil that surrounds several of our fruit trees—a perennial gift of blackbirds and sparrows who pause from flight on the galvanized cages that protect these young trees.

I have read of particular tumbleweeds, Russian thistle and tumble pigweed, that can act as agents of seed dispersal. The mother plant goes on a roll to reproduce. I would gladly trade in our weed nemesis, redroot pigweed, for the tumbling version, if I could be assured that the seed would scatter only after an extended exodus. The seed of pigweed, at least the species that we are saddled with, is known to live in the soil, if undisturbed, for more than 40 years.

Wolfgang Stuppy, the seed morphologist at the Millennium Seed Bank Partnership, part of London's Kew Royal Botanic Gardens, tells of the ivy-leafed toadflax that has evolved to bury its own fruits in crevices and fissures of rock. The flower of toadflax prefers to turn toward the sun, but after pollination, the plant searches for the shade in between stone to deposit its fruit. When fully mature, seeds are released into the elemental darkness they rely on for germination. I have witnessed the success of this innovative dispersal, having lived in England for many years. Especially in Cornwall, the "stone hedges" are a profusion of vegetation and color. These hedges that line the narrow roads almost have the appearance of garden planters, though the key to such abundance and innovation is not the result of human hands but the ingenuity of seed plants.

The artist Basia Irland has used her hands to sculpt a response to the loss of biodiversity,[5] to the barriers inhibiting seed dispersal, and to the need to conserve community. She carves ice "books," with seeds embedded as text, to release into rivers and brooks to restore the ecological balance. The ice books are stunningly beautiful—both visually and as a way to awaken our consciousness to interspecies connection. Working with ecologists, biologists, and students, Irland identifies the most beneficial plant species for each riparian zone—cottonwood, maple, lavender, or mountain mahogany—then she "composes" her book to carry the chosen seeds on the chosen water. We may call it restorative art, but it is also functional art. The seed text she has sculpted will merge with the watershed; after release into the river, the books of ice melt, releasing seeds to the riverbank. Her art is absorbed into habitat, slowing erosion, building river-bank soil, and providing shelter for micro- and macroorganisms. She enlists and educates an emerging class of "watershed citizens." Nature writer and philosopher Kathleen Dean Moore adds this lyric to the sculptor's book:

> A seed is a conveyance system for information. It is words taken wing—words written in the language of adenine, cytosine, guanine, thymine, ancient instructions clasped between hard covers, everything needed to carry a story to a new place where it

---

5   The ongoing loss of biodiversity is caused by population growth and expansion, by habitat destruction, and through the widespread use of industrial agriculture techniques (monocultures).

can take root. Long before writers figured it out, seed-bearing plants had found a way to convey to the next generation wisdom accumulated over millions of years. A samara is wisdom with ailerons. A dryas seed is a set of instructions with hair as wild as Einstein's. A dandelion seed is an epic on a parachute. A sandbur seed is a poem stuck to a sock. An elm seed is a prayer book: This way is life. This way is rootedness.

Words are the daughters of earth, I hear Samuel Johnson, the author of the innovative dictionary, say. Beside Basia, in spirit, he coaches the book of ice, encoded with cottonwood, into the flowing river.

The Great Unknown Power, the Grandfather Power,
Unknowingly, was part of the sun and the sun was part of him.
Unknowingly was seen-unseen and had many forms. He spoke:
"Ho! Aho! Now it is done. This is the Great Way of the Great Spirit
talking." And of the earth he said: "This will be my seat. This will
be my backrest." In the earth he planted the seed of life, a planting that
took half a million eons of creation time.

—told by Leonard Crow Dog on Grass Mountain at Rosebud
Indian Reservation, South Dakota; recorded by Richard Erdoes

CHAPTER 4

Acrobatic Time

MY OED (*OXFORD ENGLISH DICTIONARY*) defines *mythomaniac* as "one who is 'mad on' myths." I am not a card-carrying member of that club, though for me the word does have a magical resonance, and I find the popular usage of *myth*—to describe falsehood—to be somewhat limiting. The first definition of *myth* listed in the OED reads: "A purely fictitious narrative. . . . "

This is certainly the predominant meaning today, and I encounter it often in the daily newspaper or when I am researching agricultural issues:

> "The American dream has become a myth."
>
> "It is a modern myth that DNA contains all the information necessary to produce that organism."
>
> "The struggle against these developmentalist myths about modernization is not antiscientific."
>
> "The myth of increasing yields is the most common justification for introducing genetically engineered crops in agriculture."

The Fatal Harvest Reader, a recent book dedicated to "the agrarian mind . . . and to wildness" that covers the excesses and failures of conventional farming, opens with a bold exposé of the "seven deadly myths of industrial agriculture."[1] In his book *Seeds of the Earth*, published in 1979, Pat Roy Mooney also uses the word *myth* to argue for maintaining biodiversity: "The key to mobilizing the political 'will' required to protect the world's genetic base lies in understanding some major myths: a. the myth that the 'population explosion' threatens our food

---

1   For example: "Myth One—Industrial Agriculture Will Feed the World. The Truth—World hunger is not created by lack of food but by poverty and landlessness, which deny people access to food. Industrial agriculture actually increases hunger by raising the cost of farming, by forcing tens of millions of farmers off the land, and by growing primarily high-profit export and luxury crops."

resources and makes necessary the kind of draconian development strategies evidenced by the Green Revolution. . . . "

More recently, through the Small Planet Institute, Anna Lappé, whose energy and resolve is exemplary, has created a forum to expose the aggressive advertising of agribusiness. She has named it "Food MythBusters." The intention is admirable—to keep pace with the often "purely fictitious narratives" of agribusiness and to inspire activism.

I appreciate the potency of the word in these contexts and I agree in principle with the authors: Those who defend an industrialized approach to food systems follow an unswerving ideology more than they do fact. But I would like to explore and to revive the deeper, richer meaning of mythology as proposed by the Italian writer Italo Calvino: "Myth is the hidden part of every story, the buried part, the region that is still unexplored because there are as yet no words to enable us to get there." For me, the hidden part of nature's story, the buried part, is the life force embedded in the embryo of a seed.

"We need myth," advises Karen Armstrong, the historian of religions and author of *A Short History of Myth*, "that helps us to venerate the earth as sacred once again, instead of merely using it as a 'resource.'" She reminds us that as humans we "have always been mythmakers." With our rational minds, we are capable of extraordinary achievements in restructuring our physical environment, but it is our imagination that enables us to create art and music and to tell stories. The great German Indologist Heinrich Zimmer, writing of the myths and sagas of the Celts, said: "The magic of love and the senses, the power of nature and the unconscious, are a more imperious force than will and renunciation, consciousness and reason." We have to take a leap to enter the world of mythology, but this is a leap that can enhance our connection with the natural world, which is the wellspring of story.

Armstrong opines that our "mythical tradition has fallen into disrepute" and has yielded to the dominance of scientific reason. She uses the words *mythos* and *logos* to describe the mind's two principal systems of thought. Spencer Wells, in his book *Pandora's Seed*, which draws on Armstrong's writings, defines these terms:

> Mythos is a mystical way of viewing the world, one preoccupied with received meanings about significant events. Logos—the word provides the Greek root for the word "logic"—is the realm of rationalism, science, and Enlightenment thought. Mythos is about accepting the spiritual aspects of the world, while logos is concerned with questioning and understanding. For thousands of years, human societies have incorporated aspects of both, but in the past few centuries logos has come to the fore.

I am eager to spend some time visiting the mythos of seeds, which rise out of our interconnectedness with plants. I admit I have a bias here, as I agree with poet Stanley Kunitz's assertion (offered in his 80th decade) that "Poetry . . . is ultimately mythology, the telling of the stories of the soul. . . . [T]hese stories recount the soul's passage through the valley of this life—that is to say, its adventure in time, in history."

I do not mean to wholly dismiss the technological achievements of our present era or to reject our cultural preference for logos. Instead, I want to heal the rift we have created between ourselves and the natural world. Mythic thought takes its raw material from nature, as Claude Lévi-Strauss points out, so perhaps if we search for what is hidden in the memory of things—such as seeds—we may realize our inheritance as citizens of nature and not masters of it.

꒰꒱

BRAHMA, THE CREATOR god of Hindu mythology, is said to sit in meditation upon a lotus flower afloat on the cosmic sea. The flower grows from the navel of Vishnu, a sleeping god, who dreams the universe. When Brahma opens his eyes, a world comes into being; when his eyes close, a world goes out of being. Each Brahma lives for 4,320,000 years; when he dies, the lotus flower retracts for an incalculable, timeless second. Another lotus appears from Vishnu, who sleeps atop the cosmic serpent, Ananta, suspended above the abyss of the cosmic ocean, and another Brahma comes into existence seated on the flower. Each world so created is governed by an Indra, himself a god, unaware that to be a god is not a guarantee of immortality. What arises from the unknown waters, and from the one who dreams the universe, is the flower of divine grace, a flower that engenders seeds to enliven and color the temporal world. Thus, the lotus flower is entwined with creation and cessation.

Qualities of visible and invisible creation, part of the material of myths, permeate an embryonic plant within a seed casing. The story of plant and human life arising from a single cell afloat in the oceans of the earth may be read as a chapter in evolution but also as cosmic mythology. Remember, as humans we have always been mythmakers. According to one of the foremost authorities on mythology, Joseph Campbell, "as dreams arise from an inward world unknown to waking consciousness, so do myths . . . so indeed does life." Carl Jung, one of the innovators of psychoanalysis, attributed the power of myth to archetypes, the inner architecture of mythology. "Archetypes are many things," writes the Jungian scholar Jean Houston, "primal forms, codings of the deep unconscious, constellations of psychic energy, patterns of relationship. Our ancestors saw them in the heavens, prayed to them as Mother Earth, Father Ocean, Sister Wind. They were the great relatives from whom we were derived, and they gave us not only our existence, but also prompted our stories, elicited our moral order. Later, they became personified in mythic characters and their stories. . . . "

Before the characters of myths—gods, other immortals, fantastic creatures—were viewed as comic or irrelevant, these characters were understood as the embodiment of invisible forces, and our ancestors were nourished (and governed) by the stories. In the creative interplay between humans, gods, and the natural world, we were intimately connected with seasons, cycles, the fertility of soils, the sources of our food. Myth, far from illustrating a flight from reality, actually served to ground us within a cosmic and earthly order, grasped intuitively. Mythology has always transported us beyond the limits of our daily experience into a realm we cannot name. Karen Armstrong notes: "Unless it is encountered as part of a process of regeneration, of death and rebirth, mythology makes no sense."

Heinrich Zimmer, commenting on a 13th-century Hindu monument he encountered in the Musée Guimet in Paris, uses some language I am drawn to. He describes the stone sculpture—representing the mythology of Shiva, Brahma, and Vishnu—not as something static, but as an example of "the phenomenon of the growing, or expanding form." For Zimmer, the monument depicts a dynamic view of life that is one of the fundamental conceptions of Hinduism: Solid stone, sculpted by man, is animated; it is seen to be unfolding, expanding. In the world of plants, if viewed superficially, a seed also may be viewed as a solid object, but in reality a seed embodies a growing, expanding form. In the case of a stone sculpture, we must imagine movement and growth, but place a seed into the earth and you will witness a basic phenomenon of nature.

꙳

WHEN HUMANS FIRST began to practice agriculture, the experience of planting and harvesting was infused with the sacramental. Stories of crop cycles and soil, and the rituals practiced in celebration or out of fear, served to explain seasonal change and the vagaries of weather and of heavenly bodies. In the creation myths of Eurasia and the Americas, man and woman were depicted as growing into life out of the soil, as plants do (Genesis 2:7: "And the Lord God formed man of the dust of the ground, and breathed into his nostrils the breath of life; and man became a living soul."). Rising out of the darkness of earth, like seedlings, human life germinated into the light world above. Multiple mythologies tell of heroic journeys as a descent into the earth to bring life back again out of darkness, out of subterranean depths.

In his influential study of magic and religion, *The Golden Bough*, the Scottish social anthropologist Sir James George Frazer devotes several chapters to the Eastern Mediterranean gods of vegetation—Osiris, Tammuz, Adonis, and Attis. The ancient Egyptian tale of Osiris is emblematic of the mythological relationship between humans, gods, and soil fertility. As the first king of Egypt, Osiris introduces his people to the practice of agriculture; out of jealousy, his brother Seth dismembers him and scatters his body throughout the land of Egypt. In search of her lover Osiris, Isis wanders the land and eventually gathers together the scattered parts of his body and buries him in the earth. To honor her devotion, the gods resurrect Osiris as lord of the underworld. His resurrection symbolizes the rebirth of grain each year; when seeds germinate, Osiris is said to rise from the dead.

In a related mythology, the Mesopotamian goddess Inanna descends into the underworld in an attempt to dethrone her sister, Ereshkigal, Queen of Hell, Mistress of Life. Inanna is defeated and put to death, though she is sent back to the land of the living by other gods. She returns to discover her husband, Dumuzi, on her throne. Angered, Inanna banishes Dumuzi to the underworld with his sister, Geshtinanna. Each year he returns in the spring to Inanna, and with him the seasons change, the earth warms and revives. Embodying the primary natural cycles of vegetative growth and decay, Inanna and Dumuzi reappear in other cultures as Ishtar or Astarte, Tammuz or Adonis. They are seed-beings of a fecund and fickle earth.

There are many Native American mythologies that address one of the world's staple food crops, maize. The name itself has numerous antecedents, but Carl Linnaeus chose to classify this plant as *Zea mays* (cause of life, our mother). Though it is believed that maize, or corn as we know it, is descended from a wild grass indigenous to Mexico, teosinte, the genetic link between the corn we now cultivate and its wild cousins is not fully established. The following story is an attempt to explain the unknown origin of this invaluable food, in mythological terminology.

One of the first humans to live on Turtle Island was tired of digging for roots, so he lay down to dream in the grass of the prairie. A beautiful woman with long, flowing hair came to him and spoke: "If you follow my words, I will remain with you always." She gathered some sticks and rubbed them together to make a fire in the grass. "At sunset, you must drag me by my hair over the hot coals on the ground." When he did so, in each place that her body had passed over, a grasslike plant germinated from the earth. From that time on, he would dig, not for roots, but to plant the seeds of maize—his people had been given a new source of food.

※

ONE WINTER IN my early years as a gardener, between the cycles of harvest and planting, I came face-to-face with the inner architecture of these agricultural mythologies. I was no longer reading about characters in a story; instead, I was a participant within a story animated by the power of tides, wind, and rain. A quality I cannot define—though I could feel it bodily, as a child feels in sympathy with his or her own parent—elevated the power of the elements to the realm of mythos.

A myth usually describes an event that has allegedly occurred in time, though the essence of the story, the hidden meaning, transcent our human time. If we listen to what a myth has to say, the chaos of daily life may fade from prominence, and one may feel the "still point"[2] beyond ordinary thought or experience. Myth can offer us insight, prepare us for the unknown or undesired, or serve as a guide to the deeper meaning of life contemplated for millennia and explained as the realm of the sacred and the divine. Mythology can teach us to take a different look at the world—even the world of contemporary man where the threat of global conflict and of

---

2   From "Quartet No. 1: Burnt Norton," *Four Quartets*, T. S. Eliot: "At the still point of the turning world . . . at the still point, there the dance is, but neither arrest nor movement."

species extinction can appear as the sole reality. The "perennial philosophy," as myth has been called, can awaken us to compassion and understanding rather than oppress us with the significant pressures of modern existence that often lead to alienation, ennui, or despair.

That winter, I lived on a hillside in Cornwall with my wife and our first child, facing the waters of Mount's Bay, and we were often in the grasp of the elements; the experience was always unpredictable and often exhilarating and transformative. That harsh winter, when we experienced the loss of our second child, a girl, stillborn, my studio became a kind of sanctuary where *I was permitted to return*[3] for a short time each day. I kept a notebook open on the desk I had built, the surface covered with books and packets of seeds, and daily I was drawn to the white page. I had made the desk out of a slab of mahogany shipped from Burma to the village, located on Mount's Bay, for use by the artist Pog Yglesias. The studio had been built on the steep cliffside so that Pog, a sculptor drawn to Cornwall from London, could complete her carving of Christ.[4] One of the founders of the Mousehole Wild

---

3   The phrase is borrowed from a poem by Robert Duncan: "Often I am permitted to return to a meadow . . . "

4   Bernard Walke, "A Good Man Who Could Never Be Dull" (as the subtitle reads), author of a fine book, *Twenty Years at St. Hilary*, commissioned Pog to carve the figure of Christ. Bernard's wife was an artist, and, in a rare move for an Anglican vicar, he decorated the church with artwork from a variety of visiting and local artisans. The spire dates from the 13th century, a landmark for ships entering Mount's Bay. (A disclosure: My wife, Megan, and I were married there in 1982—we were granted permission by the Archbishop of Canterbury, and the document we received bears the seal of King Henry VIII.)

Bird Hospital and Sanctuary, she shared the studio with jackdaws. The silent space of the studio room was resonant with what I would have to name as the song of mythos.

I may have been searching for a resolution to one of the more complex dilemmas of existence. My intention, daily, was to transfer a word or a phrase onto paper, perhaps to cover the wound. I heard the words of Emily Dickinson as a haunting refrain: "My life closed twice before its close . . . " Overwhelmed by memory—the lifeless shape of our lost child—and by feelings, I found solace in poetry, the empty page on the solid desk, something I could connect with. "Reserve your rage for the gods," my teacher wrote; I did, and voiced it all to the Cornish winds. As the season turned, and narcissi began to emerge in the cliffside meadows, these words became part of a cycle, "December Songs":

I know the wind today
is a wild river.

Woman bows to woman,
shoulder to shoulder.
The moss green gleam on wood
singles a fury of wings:

Plenitude, relation of all things.

Out of the universal ocean, from beneath the lotus and the serpent, the words to accompany my story unfolded and took shape, like leaves on a stem. Perhaps I searched the book-shelves to find the story of Demeter and Persephone—I don't remember. But the story surfaced, rising from a seed I uncovered on the Cornish Penwith peninsula.

Demeter, like Isis, is the goddess of agricultural fertility and rebirth; she adores her daughter by Zeus, the beautiful Persephone. When the god of the underworld, Hades, is singled out by the goddess of love, Aphrodite, he is pierced with a passion for Persephone. He sweeps her up as she gathers flowers, and by the shore of Lake Cyane in Sicily, he opens the ground to carry Demeter's child into the earth. Demeter, mad with grief, searches the surface of the earth for her lost child. She allows crops to fail and fertility to suffer, and she causes animals to become infertile; famine rules the land. When she finally learns of Persephone's capture from the river nymph Arethusa, Demeter appeals to Zeus. Zeus negotiates for the return of his child, provided she has not eaten during her stay in the underworld. But Persephone, observed by the gardener Ascalaphus, has swallowed seven seeds of the pomegranate tree.

According to this myth, earthly seasons derive from a compromise agreed upon by the gods. Persephone is ordered to dwell 6 months underground with Hades, and the land above will remain barren; then she is allowed to return to her mother for 6 months, bringing fertility with her, and the promise of harvests.

In a barren season, as vegetation was dormant in the cliff meadows that I turned and planted in spring and summer, I could imagine an open passage into the earth, descending into the granite caves just under the soil I cultivated, and I wrote:

> *She has gone down*
> *into dark earth*
> *to taste the fruit.*

These words, part of *December Songs*, were timed to the winter melodies heard on that hillside—powerful winds, repetitive rain on a slate roof, cries of gulls, branches tapping on granite:

> *Red ivy seeds rain and split*
> *on dark earth—winter beads on stone.*
> *Light binds acrobatic time.*

> *The agitation of wind—*
> *mound moves, and limb, and cone.*
> *Look at that great space of birds!*

Poetry, as Stanley Kunitz insisted, is ultimately mythology, though it is often grounded in an earthly order. The earth turns around the sun; the seasons change. Time is an elusive, acrobatic element we cannot control. The unseen and the unknown have always sparked creation, as the embryonic life within a seed reveals itself in a plant.

When asked the question "And isn't mythology the story of one song?" Joseph Campbell replied, "Mythology *is* the song!"

*The start of every journey needs a river
and the willingness to go to a wild place.*

—Peter Forbes, *The River*

CHAPTER 5

Round Seeds,
Wrinkled Seeds

I FIND IT ALMOST IMPOSSIBLE to imagine our earth as it was 4½ billion years ago, though I did have a glimpse into that time while river rafting through the old rock landscape of Utah's Dinosaur National Monument a couple of years ago. We paddled and rode the wild water—the river speed was at an all-time high, courtesy of an abundant snowmelt in the Rockies—down the Yampa River, the only undammed tributary of the Colorado. At times, we looked up through layers of the most beautiful earthly variation of color—red, rust, sand, and mellow browns—to a rock ledge 600 feet above. The most spectacular formations seemed to sweep down like a solid river, under the liquid river we traveled on, only to emerge and rise on the opposite side as if to reveal the timeless voice of the earth's core. One of our guides whispered respectfully, as his paddle touched the cold water: "This folded rock is $1^{1}/_{2}$ billion years old." Incomprehensible to our thinking, perhaps, but we are feeling beings—and I, and my fellow travelers, could feel deep time in the landscape of the Yampa.

We are compelled to fast-forward billions of years from the time of these formations to consider the subject of this book— the emergence of seed plants within the earth's ecology— and then millions of years more until our ancestors arrived to participate, taking up seeds and planting them. In the age of the dinosaurs, when the Yampa was but a dream, club mosses, horsetails, and ferns dominated the landscape—the lush presence of equisetum (horsetail) along the river is still there to remind us. After the conifers arrived, it would be another 170 million years or so before the seed plants began their mysterious march onto land to add color and an astonishing variety of form to the green world. Travel on another 119 million years and we (Homo sapiens) arrived to share in this explosive episode of earth-given diversity by the angiosperms. Throughout those millions of years, plants became voyagers; they interpenetrated rock, clay, silt, and sand; and refined their ability to pass on inheritance. Only in the last 10,000 years have we begun to domesticate plants, to select and save seeds with some determination and deliberation. And only in the last 200 years have we come to some understanding of the purpose of pollen, and thus how plants reproduce.

It was an Augustinian friar, apparently way ahead of his time, working in a garden in Brno, who first identified some of the basic laws of plant inheritance.

Gregor Mendel was born on a farm in Silesia, in what is currently the Czech Republic, on July 20, 1822. Early in life, he was recognized as a gifted student. He studied at the Philosophical Institute of the University of Olmütz and at the University of Vienna before taking up residence in the Brno monastery, where he would become a gardener with a purpose. He possessed an active mind—he studied astronomy and meteorology, he corresponded with others regarding the sciences, and he was a dedicated beekeeper. Oddly, his work was almost unknown during his lifetime, though his experiments in the pea patch may make him one of the most gifted amateurs of all time.

From 1856 to 1863, Mendel was hard at work with his chosen legume, *Pisum sativum*, the garden pea. He planted, cross-pollinated, selected, crossed again,[1] and counted. In fact, he tested more than 29,000 pea plants in that time, evaluating for a number of traits—height, the color of flowers, and the form of fruits and seeds (round, angular, or wrinkled). And what was he searching for? In his own words, he hoped to find "a generally valid law for the formation and development of hybrids." A modest goal, though his discovery opened the door to a science he could not have imagined.

---

1   *Crossing*, or cross-pollinating, is the term used to describe the transfer of pollen from the anther (male) of one plant to the stigma (female) of another. It is also defined as a fertilization involving two different parents; to make or cause such a fertilization. Mendel was the cause agent in his garden—he selected which plants to cross.

We should pause here to more clearly define the term *hybrid*, which has come to have a much broader application than in the era of Mendel's experiments. In general terms, a hybrid is the offspring of a cross between parents of different genetic types (two varieties of a plant or animal). In narrow agricultural terms—and this is the most common meaning since the 1930s—a hybrid is a variety resulting from the cross of two inbred lines. An inbred line is the result of the repeated crossing of two varieties of the same species.

A hybrid that is the offspring of two inbred parents often exhibits something called hybrid vigor, making such a variety appealing to a gardener or farmer. However, the seed produced by that same hybrid, once saved and replanted, will not breed true; the appearance and the performance of the second-generation plant are unpredictable. Thus, to obtain the desired results of the original hybrid variety, the gardener must repurchase seed from the seed purveyor. By contrast, "open-pollinated" varieties are considered to be stable—they are the result of natural pollination in field conditions and so their seed will breed true. So if I save the seed of 'Moskvich', a handsome open-pollinated tomato with exquisite taste, and replant that seed the following year, I am confident that the 'Moskvich' offspring will look and taste the same as the parent. If I save the seed of 'Sungold', a delicious hybrid cherry tomato, and replant it, I am unsure of the fruit that will appear in the following season. And the tomato enthusiasts I know, who adore this hybrid for its exquisite flavor, could be devastated.

The value of hybrids and their role in conventional and

organic agriculture have been hotly debated for decades. Michael Pollan has called F1 hybrids "the road on which capitalism invades the garden and farm"—and the widespread use of hybrids in commodity crops (corn, wheat, soybeans) has resulted in some real disasters.[2]

Hybrid seeds are not inherently evil, and more often than not our community farm members are surprised to hear that we purchase and sow hybrid as well as OP seed. After all, when plants in the wild cross-pollinate, their progeny can also be labeled as hybrids—in this case, the result of a natural process. However, when hybrids are increasingly substituted for OP varieties and used on an industrial scale, the result is an expansion of monoculture, and genetic diversity is further diminished. This spring, as I was preparing to order seeds for the coming season, I read a blog post titled "Inside Hybrid Seeds" by the founder of High Mowing Organic Seeds, Tom Stearns, that addresses this issue.

> The crossing—and subsequent natural selection— that happens with the new combination of genes has been the core aspect of how natural diversity has developed on our planet. . . . As people started to cross plants more intentionally, a new aspect of hybrids developed in the economic realm. For nearly a century seed companies have been using the natural

---

2   The *F* stands for *filial*—the generations that follow from a controlled cross-pollination (rather than a cross occurring in the wild). Hybrid garden varieties are identified as F1.

combining ability of two different varieties to develop new varieties to meet the needs of farmers and gardeners. They learned that you could do things with hybrids faster or differently than you could with open-pollinated varieties. . . . But early on some seed companies also realized that with hybrids their customers could no longer save their own seeds and get the same varieties the following year. Customers became dependent on the seed companies in a different way.

It is not likely that the diligent friar working in his garden in Brno had any idea of where his discovery would lead, though he did admit, "My scientific studies have afforded me great gratification, and I am convinced that it will not be long before the whole world acknowledges the results of my work."

However, Mendel did not live to witness that acknowledgment. In 1866, he published a paper on the results of his scientific husbandry, "Experiments of Plant Hybrids," and his paper received almost no notice. It is commonly thought that Mendel was answering questions that no one else was yet asking. In fact, he had no specific word to describe what he found—the invisible thing, the something, that determined heredity in his pea plants. Mendel, as he counted and recounted violet and white flowers, tall and short plants,[3] referred to

---

3   Mendel used letters to describe different alleles of genes that were present in his crosses. He used *T* for the tall allele, and *t* for the short allele, for instance. An allele is an alternate version of a gene—an allele for tallness, an allele for shortness; an allele for blue eyes, an allele for brown eyes, etc.

dominant and recessive traits to describe the forces that drove heredity and created variety; he referred to elements and to factors because no other terminology existed to name what his experiments were revealing to him. As an explorer in the plant world, he was face-to-face with the *force that evolves a form*. He knew that these factors existed in plants as a unit of heredity, but he could not name the substance. Instead, he chose to name two laws: the Law of Segregation and the Law of Independent Assortment. Though his theories eventually were proven to be correct, in his own time there was as yet no understanding or mechanism to evaluate his conclusions.

In 1900, within a few months of one another, three botanists—Hugo de Vries, Carl Correns, and Erich von Tschermak—each published articles that corroborated Mendel's work. They had reached their conclusions independently, and only upon researching the existing literature did these three rediscover Mendel's achievement of 40 years earlier. Mendel's discoveries finally greeted the world with something like the power of revelation.

Mendel's genius was to reveal an insight into the laws of nature by looking clearly at traits within a particular species that did not show great variance. The mathematical approach to biological phenomena that characterized Mendel's method was entirely new (his calculations went a step further than those of Charles Darwin, who also experimented with plant hybrids at roughly the same time), and this is what allowed him to formulate his laws. Plant breeder and biologist Carol Deppe writes that Mendel "had exactly zero impact on the field" (of genetics), but that the phenomenon of independent assortment, the conclusion formulated by Mendel after counting specific traits throughout generations, over and over, "forms the basis for much of plant breeding."[4] As such, his discoveries form a basis for our scientific understanding of plants and for the production and distribution of seeds. The entire field of genetics would continue to unfold in the 50 years following the rediscovery of Mendel's careful work, including the identification of chromosomes ("colored bodies," from the Greek), genes, and eventually the discovery of DNA.

Mendel's experiments in his garden have had far-reaching implications across a variety of disciplines, and quite beyond his original intentions. In *Genetics and the Manipulation of Life*, Craig Holdrege, a biologist and director of the Nature Institute in Ghent, New York, observes that as a result of Mendel's experiments, the field of genetics "took the path that led to the clearest interpretation." But that path has led us away from a central relationship we now need to cultivate—the *interrelationship* between our species and all others. Holdrege's argument is that thinking only of the object, while useful for predicting and arriving at results, "does not penetrate to the more open, developmental potential of the organism, which lets it interact with its environment and change according to changing circumstances." Discussions of Mendelian genetics tend to ignore the reductive elements of his experiments. A more fluid thinking process would consider the "plant's potential to

---

4   From Carol Deppe: "Independent assortment allows the breeder to cross two varieties that differ in a number of ways and to derive new varieties with different combinations of the characteristics of the parents, or even completely new phenotypes."

change its form and structure in accordance with changing environmental conditions." This is exactly what a farmer or gardener, with open eyes, will witness every day in the pea patch, or in the rows of potato plants, or by following the trailing squash vines to the blossom end.

THE EXISTENCE OF the cell nucleus was discovered in the 1830s, and in 1842, a Swiss botanist named Karl Wilhelm von Nägeli observed that the nucleus would sometimes congeal into small rod-shaped bodies—later to be named chromosomes, coincidentally in the same year that Mendel died, 1884. When Mendel spoke of "factors" and of "elements," he was well aware of something behind the traits that he observed in multiple generations. The factor that he imagined would later be named a *gene*. But remember, as Mendel searched for an abstraction, the "something" he deduced to be there, he was searching in the garden—with his own hands, this monk was cross-pollinating plants. His living lab was rooted in the soil.

When we proceed from Mendel's garden into the technological explosion that marked the 20th century, we are forced to comprehend ever more complex and evolving terminology in order to approach some understanding of nature's precise mechanisms. As a farmer, I am always open to the things I cannot see, because I know that the ultimate health of our soils is determined by billions of invisible microorganisms that continuously re-create a living environment. I have no need to investigate the genetic parentage of crops in fields that border our farm, but, increasingly, other farmers do, to protect the genetic integrity of their field crops. A familiarity with some of the history and terminology that has influenced the conceptual framework of our food system is increasingly essential for those who grow our food and for those who consume it.

Although we commonly think of a gene as an object, our understanding of this "functional segment of DNA associated with the synthesis of a protein" is still evolving. I wonder what Mendel would have thought of this, written in 1991 by Ernst Peter Fischer and translated by Holdrege:

> There are, therefore, no genes! There are—at least in the cells of higher organisms—only pieces of genes, which the cell can use when it makes proteins. A gene is by no means a molecule that exists in the cell. Rather, a gene is a task that a cell has to accomplish. Genes don't exist; they are always becoming.

When I read this, I am keenly aware of my kinship with the evolutionary process that has given us flowering plants and seeds. We are newcomers on our planet, though daily we neglect this essential information, until by chance we are face-to-face with folded rock flowing under an ancient river. At one time, life existed only as single cells. Seeds, carrying the germ of life within, the endosperm, arose from these cells—retaining their nucleic memory, but with an urgency to grow and to become.

The relationship between language and landscape is a marriage
of sound and form, an oral geography, a sensual topography,
what draws us to a place and keeps us there.
Where we live is at the center of how we speak.

—Terry Tempest Williams, "Red"

CHAPTER 6

"What a Find!"

As LAST AUTUMN WAS DRAWING to a close, with the pepper
plants still full of fruit, I went out to the far field to collect the
best of the small gems we know as 'Sweet Fingerlings'. We
were introduced to this variety originally by Ken Ettlinger, a
well-known local seed saver who teaches biology at a nearby
college. "Given" to us would be the more appropriate phrase,
used in similar fashion by musicians to acknowledge the way
in which a song or melody is passed along: "This was given to
me by John Gawler at the Maine fiddlers' camp in the summer
of 2001."

We met Ken when our farm participated in the Public Seed Initiative and the offshoot of this, the Organic Seed Partnership, programs jointly sponsored by NOFA-NY (Northeast Organic Farming Association of New York) and our land grant university, Cornell. He hosted a field day at his magical garden in Flanders, a hamlet near the mouth of the meandering Peconic River—one of the few rivers that empty into the great bay near the tip of Long Island. It was there that we tasted the miniature pepper, shaped more like a jalapeño but sweet beyond belief. Ken gave us some seeds when we expressed an interest, and we saved the offspring for a number of years, until late blight disease arrived in the tomatoes in 2009 and we lost track of all other crops. Given some time, and in the absence of a crisis, it is not difficult to save pepper seeds. Identify a plant with a robust nature, leave a few fruits—in this case, well-formed peppers—to ripen, and harvest the chosen peppers before the threat of a frost. Remove the seeds from the cavity of the fruit and lay them out to dry in a well-ventilated place. When fully dried (the time period will vary according to humidity), store in a packet, a jar, or a plastic vial in a cool place (36 to 40 degrees) until planting season comes around.

I wrote to Ken in the spring of 2011 with a request (please!) for a few seeds of 'Sweet Fingerling' peppers. What I received in the mail a week later was startling—though perhaps unsurprising for a gardener enamored with plants and their progeny. Ken had supplied us with the seeds of the variety we desired—a challenge in itself because the 'Fingerling' plant produces very few seeds per fruit—but he also gifted us with more than 20 varieties of vegetables that he had tended, collected, and saved for years, varieties familiar with Long Island soil. When a gardener or farmer sows seed, there is always a voice—centuries old—attending the sowing: "Pass it on. . . . " Farmer to farmer, gardener to gardener, that voice rings clear (observe the 500 pages of the annual Seed Savers Exchange Yearbook, listing 12,495 varieties). However, once you introduce the middlemen, the corporations focused first on profit margin, then seed saving as a functional art is lost. The job of maintaining the diversity of our seed supply now rests predominantly in the hands of committed amateur seeds savers (we will return to this subject in a later chapter).

Ken's gift of seeds included this note: "Here are some seeds of selections in my gardens. Included are the Sweet Fingerling peppers. They are so frugal in seed production. There are reds, oranges, and yellows, and also a large red selection that showed up a few years ago." The commentary accompanying each packet of seeds was flavorful and informative, offering a window into the thought process of a seed saver:

> **LI Sweet.** This is a productive and early pepper that is a dehybridization of Gypsy. A six year journey. It is mostly a yellow-green immature pepper still with some variability.
>
> **HotCha.** Thick flesh hot pepper with uses similar to jalapeño although it has a large wedge shape. Better adapted to Long Island.

Summer Bush Squash: Zephyr Crosses. Visually surprising group of long summer squash; diverse, excellent flavor. Zephyr dehybridizes back to its bumpy yellow squash pedigree but its crosses with zucchini produce long smooth dark green zucchini and a variety of intermediate banded and striped kinds.[1]

Dragon. Wide long cone shaped attractive purple and green pepper changes to red. Sweet when immature; hot as it ripens. . . . Dehybridization of a hot pepper from China.

Because of Ken's patient seed saving and breeding experiments over many years, the 'Sweet Fingerling' peppers will live on in our fields. When he first presented us with this variety, Ken explained that we would encounter both red and yellow peppers when the plants came to fruit. If we so desired, we could "stabilize" the variety by saving seed from one or the other; to date, given the peppers' exquisite sweetness, I prefer to encourage the range of color, and with it the element of surprise. Ken writes: "I maintain many of the OSP [Organic Seed Partnership] varieties and I am distributing some of those through Seed Savers and other avenues so they keep going around." At

---

1   Ken is a breeder/seed saver who advocates working with hybrids as well as open-pollinated plants. To "dehybridize" is often a multiyear process of selecting and saving seed to return to a previous parentage—with a desired trait in mind and a passion for one specific shape or color or taste.

our community farm, we agree with this philosophy: Let's keep
them going around.

࿔

IT IS SAID that the great Russian plant explorer Nikolai Ivanovich
Vavilov could recite entire books by Pushkin, word for word (I
would have loved to witness this feat). He could speak English,
German, French, Spanish, Latin, Farsi, and Turkic. He is
quoted as having said, "Life is short, we must hurry." Though
he did manage to hurry—and to record a beautiful, creative
legacy for more than 50 years—his life was cut short at the
whim of Stalin in 1943. Spencer Wells observes that Vavilov led
a life "that deserves to be a feature film," for the Russian inno-
vator was convicted and imprisoned, and he eventually perished
in the gulag at the height of his creative work. Peter Pringle's
2009 book, *The Murder of Nikolai Vavilov*, could be the basis for
a script for that film, and Gary Paul Nabhan's *Where Our Food
Comes From: Retracing Nikolay Vavilov's Quest to End Famine*
underscores the importance of Vavilov's work and travels for our
own time. Over the course of many years, Vavilov visited and collected seeds from dozens of countries on five continents, and
Nabhan retraced the footsteps of the energetic Russian in a total
of 10 of those countries, ranging from Afghanistan to Colombia.
Nabhan writes that he has been chasing the "shadows and seeds"
of Vavilov since 1976, when a friend presented him with a 1950
edition of the horticultural journal *Chronica Botanica* containing
an account of the extraordinary botanist.

As the head of Russia's All-Union Institute of Applied Botany and New Crops—which eventually employed 20,000 workers—Vavilov was assigned to improve farm productivity in a country ravaged by war and revolution. He was a prescient thinker and advised his staff at the institute that their work was not only to collect plants for the purpose of breeding but to rescue seeds from extinction. Their mission was to search out and to recover the wild relatives of cultivated plants: Vavilov was keenly aware that in order to improve food crops, it was essential to maintain a diversity of plant resources, which preserves genetic variability and ultimately increases the likelihood that plants will adapt to climatic shifts.

Peter Pringle refers to the Russian plant explorer as a "bogatyr," a Hercules, a man of extraordinary powers.[2] He was able to fare forward on only 4 to 5 hours of sleep per night—and fare forward he did, leading 115 expeditions to a total of 64 countries. The authors of *Shattering* (Cary Fowler and Pat Roy Mooney) acknowledge Vavilov's far-sighted thinking: "The great geobotanist and ultimate plant explorer was the first to recognize genetic erosion as a global threat to food security, and to see seed keeping as a conservation strategy." The institute that he directed, and that now bears his name, in effect became the first world seed bank because of his vision and because of his strong legs—he was a born explorer. There is a revealing passage

2   I am well acquainted with the word's meaning—our most robust cultivar of *Allium sativum* var. *ophioscorodon*, a garlic variety we harvested in 2012 and replanted last autumn, is known by this expressive name, 'Bogatyr'.

from Vavilov's *Five Continents*, in which he records his experience exploring the Afghanistan/Tajikistan border:

> It seemed that we had finally passed this very difficult trail so that we could mount the horses and continue on. But suddenly from the cliff above the trail, two gigantic eagles flew out from a nest, circling on enormous wings. My horse shied and bolted, galloping along the trail and the ovring.[3] The rein was unexpectedly torn out of my hand and I had to hang on to the mane. Above my head were cliffs but below me, 1000 meters down in the deep ravine, rumbled the beautiful, blue Pyandzh, the upper reaches of one of the great rivers of Inner Asia. That is the experience, which afterwards this traveler remembers best. Such moments steel one for the rest of one's life: they prepare a scientist for all difficulties, all adversities, and everything unexpected. In this respect, my first great expedition was especially useful.

Vavilov embarked on his life's work at an extraordinary time. Gregor Mendel's theory of inheritance had just been rediscovered—Vavilov was an early defender—so the study of genetics was unfurling while the political and cultural landscape of Europe was in turmoil. He had the early support of

---

3   A rather treacherous cliffside path commonly found in Afghanistan.

Lenin, but following Lenin's death, the political landscape of Russia shifted significantly with Stalin's rise. It was almost inevitable that the curious geobotanist would clash with the conservative dictator, though it took years for the machine of the repressive state to arrest the explorer. Vavilov was searching for long-term solutions that would guarantee future food security; Stalin was demanding an immediate fix for the specter he feared, famine. "Martyr to genetic truth" is the title bestowed upon Vavilov by James Crow in a paper published by the Genetics Society of America.

Before the arrival of Vavilov on the world stage, it was assumed that agriculture as we know it was born along the Tigris and Euphrates Rivers. But this Russian plant geographer, following his research and his travels, arrived at a different theory. He identified eight centers of origin of cultivated plants, and these centers were found primarily in mountainous regions (between latitudes 20 degrees and 45 degrees north), not in river valleys. The centers Vavilov identified were located in China, India, Central Asia, the Near East, the Mediterranean, Ethiopia, and southern Mexico. He also included sites in South America—Peru, Ecuador, and Bolivia. To explore and research plant origins in the mountains was to uncover various "refugia of biological diversity." Originally, Vavilov reasoned that by locating the place of greatest genetic diversity in plants, one could identify the place of origin, though later he recognized that this was not always the case. Had he lived longer and continued his explorations, it is likely he would have amended his theory, as researchers since have done. Vavilov's maps have been redrawn, and the centers of diversity as he first imagined them are not as precisely delineated, but the "beauty, simplicity, and utility" (as the authors of *Shattering* express it) of his theory remain. There is a close correlation between the "Vavilovian centers of diversity" and what have been named the "hot spots" of biodiversity, a focus of present-day plant explorers. In Nabhan's estimation, this concept of the centers of diversity, presented in a 1926 publication, is in fact Vavilov's most significant contribution to science.

By 1940, Vavilov and his team had collected more than 250,000 plant species from five continents; these living samples were distributed and studied in more than 400 experiment stations scattered throughout Russia. To steward and transport such material many thousands of miles—from the highlands of Ethiopia to a field station in Russia—obviously required significant planning, coordination, and chutzpah. Jack Harlan, whose agronomist father hosted the Russian explorer on his trip to the United States, wrote of Vavilov: "It was his plan to collect and assemble all of the useful germplasm of all crops that had potential in the Soviet Union, to study and classify the material, and to utilize it in a national plant breeding effort."

His methodology, however, increasingly came under scrutiny by Stalin's state; while the new science of genetics was gaining followers in other countries, in Russia the discipline was doubted and dismissed. Trofim Lysenko, a horticulturist from a peasant background (which was in favor by the Communist regime), who theorized that it was possible to "train" plants by altering their environments, became Stalin's rising star. Due to crop losses and the failures of collectivization, Russian agriculture between the World Wars was in a state of crisis, and Stalin expected results in 3 years. Vavilov's intention, to bring an end to recurring famine using careful selection, would take time. Lysenko, however, promised the results that Stalin demanded—though he had no basis for his confidence—and his particular brand of pseudoscience was rewarded: He eventually usurped Vavilov's role as director of the institute.

Though in the end Vavilov was, quite literally, hunted down in his own homeland, he was welcomed by both scholars and farmers in other countries that he explored.[4] Nabhan illustrates this with the story of a 1930 lecture given by the Russian plant geographer to a full auditorium at the land grant University of Arizona in Tucson. He delivered his talk, "The Origins of Cultivated Plants," in polished English, advising his audience of the debt owed by nations such as the United States to the developing countries, the birthplace of much plant genetic diversity. Abyssinia, he noted, though a tiny country, is home to the largest selection of wheat in the world. Then, referring to the purpose of his visit to the "New World," Nabhan quotes him as saying: "It is really time to begin the discovery of America." Vavilov was challenging his audience to explore and document crops, not yet discovered, native to American soil. He expressed

---

4   As an epigraph to his book *The Murder of Nikolai Vavilov*, Peter Pringle quotes Aleksandr Solzhenitsyn: "You would have no love whatever for your country, you would have to be hostile to it, to shoot the pride of the nation—its concentrated knowledge, energy and talent! And wasn't it exactly the same . . . in the case of Nikolai Ivanovich Vavilov?"

excitement about a number of species recently recognized as being domesticated in North America: sumpweeds, chenopods, Chickasaw plums, and Sonoran panic grass. A bold statement by a visiting plant scientist, but apparently his erudition and self-assurance alerted everyone to the value of his innovative approach to crops and agriculture.

In 1940, while on an expedition in the Ukraine, Vavilov was arrested and swept away into Stalin's intricate prison system. He was convicted the following year—for spying on behalf of England, for belonging to a rightist organization, and for sabotaging Russian agriculture—reportedly after 1,700 hours of rigorous interrogation. After a show trial in 1942, Vavilov was sentenced to be shot, though after several allies interceded on his behalf, he survived for another year in the gulag. The man who had led the search for wild relatives of cultivated plants to alleviate famine perished for lack of food.

While the great plant geographer was being held in prison, Hitler's army encircled Leningrad (now St. Petersburg) in a siege that was to last for almost 3 years. Eventually, 700,000 people starved to death as a result of the siege, including a number of those who worked at Vavilov's institute—scientists and breeders—in the heart of the city. When Cary Fowler and Pat Roy Mooney visited the institute nearly 100 years after Vavilov's birth, they were shown photographs of some of those dedicated protectors of the seed. They had made the conscious choice—almost impossible for us to imagine—to guard the collection of food diversity (including rice and potatoes and pulses—members of the bean family) against animal and human invaders, rather than to consume it and to live. When asked, "Why would people starve to death surrounded by food?" the geneticist/historian of the institute replied: "They were students of Vavilov."

※

DR. KEN B. WILSON, director of the Christensen Fund (a private foundation that promotes biocultural diversity), writes in his foreword to Nabhan's *Where Our Food Comes From*: "Thus armed with the mule and the capacity to learn from the custodians of the seeds, Vavilov became the first to recognize for science that there was a geographic coherence and pattern to those ancient agricultural civilizations of mountain tribal peoples who had domesticated the thousands of plant species upon which we still depend for our food." I believe him when he speculates that Vavilov would most certainly have been fascinated by the modern science of genetics, and I wonder where his thirst for exploration would have led him in the 21st century. Wilson observes that the debates over the future of agriculture are generally oversimplified. Vavilov's science and intuition were in alignment; traditional farmers have been breeding and selecting, and thus refining the art of seed saving, for thousands of years. Genetic variation in plants provides insurance against climatic unpredictability; traditional farmers knew this and followed nature's lead. To select for resilience, a farmer will know his plants and the gifts and demands of the ecosystem. Today, private, profit-driven interests—with global aspirations—ignore the importance of place and of indigenous plant technicians, the true keepers of the seeds that matter. We should choose to maintain the freedom of those who fare forward in search of an inheritance that will promote diversity, and those who appreciate our 10,000-year heritage of wheat cultivars or the sweetness of peppers. Imprisonment for the sake of an argument is failed politics and the sign of a diminished social fabric; the man who notes the enormous wings of eagles aligns himself with the spirit of renewal and the millennial dispersal of seeds.

Time is turned by the plough, and the rose was earth.

—Osip Mandelstam

CHAPTER 7

The Memory of Life

SEED CATALOGS—INTENDED BY THEIR AUTHORS to lure readers into the next growing season—often arrive at our farm shop prior to Thanksgiving, before we have finished in the fields for the present season. I respect their intention—I know the marketplace demands an autumn publication—but I do not have the leisure to peruse a catalog in November; in recent years we continue to harvest cabbages into the week after Christmas. The catalogs remain stacked in a corner of the farm shop until the first white flakes arrive and our machinery is parked for the winter. Selecting seeds and anticipating the next season can be a delicious occupation, so I prefer to take it on in a restful mood, after the turn of the year.

Upon opening a seed catalog, one expects to read a welcoming message, perhaps a statement of purpose, and then to find page after page of varietal descriptions. The co-op seed packers from Maine, Fedco Seeds (our primary supplier), present another model. Page 5 of a recent catalog bears a long list of resources—from "Seed Saving Organizations" to "Useful Publications" to "Organic Farming Organizations." Page 7 updates the reader on a recent polemic: "Fedco joins lawsuit against Monsanto." Calling transgenic technology "the largest human-instigated biological experiment in history," this article explains: "Society stands on the precipice of forever being bound to transgenic agriculture and transgenic food. Coexistence between transgenic seed and organic seed is impossible because transgenic seed contaminates and eventually overcomes organic seed."[1]

Transgenic methodology, responsible for GMOs (genetically modified organisms), has been around for some time. Genetic engineering (GE), or genetic modification, is the act of altering organisms by the physical transfer of DNA. The altered organism, a GMO, is the result of a laboratory process by which a gene (or genes) of one species is inserted into another species. This process is fundamentally different from traditional breeding. Jack Kittredge, who has written about transgenic crops for the *Natural Farmer*, draws a graphic

---

1   This chapter is bound in another direction, but to return to the 2012 Fedco catalog: Excerpts from Franklin Delano Roosevelt's fireside chats—radio addresses delivered to the American people during his presidency—are scattered throughout the remaining 145 pages of the catalog, almost as inspiring as the numerous plant varieties, from 'April in Paris' sweet pea to 'Revolution' pepper and 'Kilimanjaro' euphorbia.

analogy: "In traditional breeding one can mate a pig with another pig to get a new variety, but one cannot mate a pig with a potato or a mouse."

After the introduction of genetically engineered crops in the United States in the mid-1990s, the acreage planted yearly has multiplied exponentially; by 2010, 93 percent of US soybean acreage was planted with GMO seed, while 63 percent of corn acreage and 78 percent of cotton acreage were composed of GMO crops. This represents a far greater percentage than in many other countries, where restraint has been the rule. (As of 2013, the planting of GMO crops is largely banned in the 28-nation European Union.) From the very start, the subject of genetic engineering and GMOs in agriculture has been contentious (I'm being polite). In *First the Seed*, Jack Ralph Kloppenburg Jr. refers to the "seed wars" of the 1980s—and obviously the skirmishes have intensified. Many of the battles are now being fought in the courts (including the US Supreme Court), and farmers are pitted directly against global corporations. Individual counties (in California) have banned the use of GMO crops, GMO-Free activists are aggressively campaigning throughout this country and worldwide, and biotech corporations are suing one another for greater market control. The question is, how did we arrive here?

※

LET US GO BACK to the era of a war that divided this nation, the Civil War, to the year 1862, in which the Morill Land Grant College Act and the Agriculture Act (establishing the United

States Department of Agriculture) were enacted. As individual farmers planted and tended crops and saved seed for the next season (in a time of war), the newly formed government department—with a mandate to improve the nation's agricultural output and to assist those who practiced it—experimented with a diversity of crops, making use of germplasm[2] collected from all over the world. Indeed, the main function of the USDA at that time was to search for plant germplasm and to distribute seeds to America's farmers, free of charge. By 1878, a third of the annual budget of the department was allocated for these two purposes. In 1897, the US government distributed 22,195,381 seed packages to growers—and because each package contained five varieties, the total distribution amounted to 1.1 billion seed packets, an impressive investment in the health of a nation's soil and husbandry. Thoreau's beautiful phrase could be used to sum up the power of that efficiency: "I have great faith in a seed . . . "

As of 1916, farmers were still an essential part of the plant breeding/seed saving community. In that year, the USDA sent out 337,442 seed packages to those tending the American soil. Breeding methods had not really changed since the founding of the nation; farmers and researchers still searched for desired traits in plants and carefully selected for what would be fruitful

2   In *Breed Your Own Vegetable Varieties,* Carol Deppe defines the term: "Like genome, the term *germplasm* refers to all the genetic material, but the context is usually broader and the emotional value is different. . . . *Genome* is a technical, unemotional term. *Germplasm,* with its associations of germ/core/life and plasm/life/blood, has mystical and spiritual overtones. We use the word *germplasm* when we care about it."

in the field. Charles Darwin offered an apt description of the basic breeding method favored by our species:

> I have seen great surprise expressed in horticultural works at the wonderful skill of gardeners, in having produced such splendid results from such poor materials; but the art has been simple, and as far as the final result is concerned, has been followed almost unconsciously. It has consisted in always cultivating the best known variety, sowing its seeds, and, when a slightly better variety chanced to appear, selecting it, and so onwards.

Yet change was on the way, and it would gradually redefine the relationships between public and private interests. Writing in 1902, at the time of the Second International Conference on Plant Breeding and Hybridization, English geneticist William Bateson, acknowledging the new breed of breeder, wrote: "He will be able to do what he wants to do instead of merely what happens to turn up." Slowly the focus of research at the land grant colleges[3] began to shift from collaboration with the farming community to the development of an agricultural science.

---

3   The land grant colleges—one created within each state of the Union—were instituted to provide technical support and assistance to farmers. Some readers will be familiar with those that have generated notoriety: Cornell University, Michigan State, Ohio State, and the University of California at Davis, to name a few. One might fairly ask who is served by the land grant colleges today, since the social/political/cultural dynamics have shifted dramatically since the passage of the 1862 act that created the land grant system.

In 1883, representatives of 34 seed companies had met in New York City to found the American Seed Trade Association (ASTA) and to promote the commercial interests of the trade. Thirty-seven years passed before ASTA finally persuaded Congress to eliminate the free distribution of seeds (1924). We now are wary of the tendency of (our) government to intrude in our private affairs, but imagine how productive the free exchange of seeds could be—in a spirit of trust—between present-day farmers, the USDA, and the land grant universities. Such was the original intention. I am tempted to say that industry got in the way, though, of course, the story is more complex. Jack Kloppenburg, commenting on the era when cooperation in the public sphere was a given, writes: "The seed industry was locked into a subordinate position to a public sector aggressive in its approach to applied science and ideologically committed to a mission of serving the farmer."

As the new science of genetics evolved, plant breeders began to experiment with more sophisticated hybridization techniques. Up until that time—for the preceding 10,000 years or more—all breeding was accomplished on the basis of mass selection. Kloppenburg makes the point that "the coupling of Darwinian and Mendelian thought . . . had a profound effect on plant breeding. . . . [P]lant breeding was becoming less of an art and more of a science." The science as first practiced was a product of our public institutions, though with the introduction of hybrid corn, the seed trade soon found a path toward significant profit. John Navazio, a plant breeder who cofounded the Organic Seed Alliance in 2003, notes that the first advantage of hybrid varieties for a seed company is that the company has instant proprietary ownership of each variety (because of their control of the inbred parents). Hybrid onions were first introduced in this country in the early 1920s, followed by the first public release of hybrid corn. In 1926, Henry A. Wallace, who later became a US vice president, founded a company with the specific purpose of selling hybrid corn, and over the next 2 decades the sale of hybrid seed—as others got into the business—rose from zero to $70 million in revenue.[4] Corn yields continued to rise in this era, and by 1965, 95 percent of US acreage was planted with hybrid corn.

Then, in 1930, Congress enacted the Plant Patent Act, which ruled that plant species that were reproduced through asexual propagation could now be privately owned, although seeds were exempt from this ruling (tubers, including potatoes, were also exempt). The ASTA pressed their case, hoping to include species that reproduced sexually (most of the agricultural crops we know), but Congress refused to acquiesce. It is fascinating, in light of the present-day debate over "genetic drift"[5] relating to GMO crops, that in 1930 the USDA opposed including plants that reproduced by seeds in the Plant Patent Act. They reasoned that enforcement of the act would be

---

4   Jack Kloppenburg uses this statement by Wallace as an epigraph to a chapter: "We hear a great deal these days about atomic energy. Yet I am convinced that historians will rank the harnessing of hybrid power as equally significant."

5   One variety of a plant, aided by the wind, will readily cross pollinate with another variety of like species planted in a nearby field, thus altering the genetic makeup of the following generation.

impossible because most seed plants, in genetic terms, are not sufficiently stable; genetic drift over generations represented an incalculable factor. Putting aside the original question—is it morally, ethically, or practically acceptable to patent living matter?—the decision makers in the USDA, in 1930, were correct.

I am not convinced that it was intentional on the part of those involved in US agriculture, horticulture, and crop science in the mid-20th century—both public and private interests—to systematically limit diversity, but this is the consequence we have inherited. Multiple factors are responsible for the present situation—the proliferation of hybrid crops, consolidation in the seed industry, the decline of public breeding programs, and mass marketing considerations, to name a few. In *Shattering*, Cary Fowler and Pat Roy Mooney include a chart listing plant varieties lost in the United States from 1903 to 1983. Of the varieties actively cultivated in 1903—from runner beans to celery—the percentage remaining in the National Seed Storage Laboratory collection is deplorably small. The range of crops lost hovers in the 80 to 100 percent range, with more than 90 percent loss most common. For example, 92.8 percent of lettuce varieties have been lost, 93.9 percent of peas, 97 percent of rutabaga, and 91 percent of watermelon.

In an article titled "Hard Times for Diversity," biologist and conservationist David Ehrenfeld comments on the cultural and economic forces that have led to this loss:

What has caused the decline of the love of diversity and is causing the decline of diversity itself is, not surprisingly, the ascendancy of its opposite: uniformity. We have abandoned our fascination with the specific, with species, in favor of a preoccupation with the general and the generalizable, with scientific laws. This is the Age of Generality, and every month that passes sees it more firmly entrenched as the official way of seeing and dealing with the world.

Increasingly, in the early to mid-20th century, public breeders and private seed companies aimed to create uniformity, to establish more predictable results—thanks in no small part to consumer demand—and to continuously promote hybrid production.

In his book *Seeds of the Earth*, first published in 1979, Pat Roy Mooney begins with a discussion of "gene-rich and gene-poor" areas, pointing to the Vavilov centers as the places of greatest genetic diversity. The gene-rich areas—those with a greater range of species, and of variety within individual species of plants—are located far from the industrialized North: Afghanistan, Indo-Burma, the Peruvian Andes, and Ethiopia, among others. He notes that prehistoric people cultivated some 500 species of plants and foraged for food among another 1,500 wild species. In contrast, our food diversity in North America is now limited to about 200 species grown by gardeners, with far fewer being cultivated on farms; only three crops—wheat, rice, and maize—supply 75 percent of our cereal needs.

As more and more variety is sacrificed, we are left with a stunning degree of food interdependence. Our food crops are primarily derived from common genetic pools originally found in the gene-rich areas of the globe. To illustrate to what degree this is true, Mooney traces the ancestry of a Canadian wheat variety, 'Marquis', from Galicia to Germany to Scotland to Canada, with the final addition of an Indian cultivar. A historic shift in crop breeding occurred sometime around World War II when, because of more sophisticated breeding techniques, commercial interests, and the early success of hybrid varieties, we began to export "uniformity." In a chapter titled "Genetic Erosion," Mooney quotes H. Garrison Wilkes:

> Suddenly in the 1970's, we are discovering Mexican farmers growing hybrid corn seed from a Midwestern seed firm, Tibetan farmers planting barley from a Scandinavian plant breeding station, and Turkish farmers planting wheat from the Mexican programme. Each of these classic areas of crop-specific genetic diversity is rapidly becoming an area of seed uniformity.

John Navazio coined the word *hybriditis* to describe the institutional fascination with hybrids. There is a marked tendency for the 20th-century breeding method of choice to usurp a crop; whereas most vegetable crops were still open-pollinated (OP)[6] into the 1960s, by the 1980s, hybriditis was the rule

---

6   Bryan Connolly defines *open-pollinated* as "any population of crop plant that breeds true when randomly mated within its own variety. Like begets like, though always with some minor variation."

rather than the exception. Maintaining an inbred line is less work for a seed company, though the result, again, is less genetic variability in circulation. Navazio comments: "Nobody has put any real energy into breeding good OPs in 50 years."

Until now, if we have hope, and stamina.

❦

THE ROOTS OF the deceptively named "Green Revolution"— in which agricultural production skyrocketed globally—can be traced back to the 1940s, when the Rockefeller Foundation helped to launch the Mexican Agricultural Program. Gradually, research stations were set up in almost every other South American country, aided by the USDA and our land grant system. Jack Kloppenburg points out that the force behind this agricultural/socioeconomic movement was fueled by "business, philanthropy, science, and politics." Dr. Norman Borlaug was awarded the Nobel Peace Prize in 1970 for his work with wheat cultivars and his discovery of a gene that gave wheat an immunity to the devastation of red stem rust. Borlaug made use of a Japanese semidwarf wheat cultivar, 'Norin 10', to incorporate a shorter wheat stalk capable of supporting a more productive seed head. In Mexico, Pakistan, and India, this variety of wheat made it possible for farmers to double their yields, or more; results such as this helped to sell a "technology" in a political climate now focused on global hegemony. The research center in Mexico was established as the International Maize and Wheat Improvement Center (CIMMYT), nicknamed "the school of the wheat apostles." Under the guidance of scientists trained here, the new, improved seed varieties were introduced in developing countries throughout the world. The Nobel Committee declared of Borlaug: "More than any other single person of this age, he has helped to provide bread for a hungry world." He is credited with saving one billion lives. At the same time, US corporations increased their profits through the marketing of commodity crops on a global scale. As the exporters of a new technology, the gene-poor nations of the North increasingly gained hegemony over the developing nations of the South.

John Seabrook, writing in the Annals of Agriculture column for the *New Yorker*, observes: "The green revolution was a complicated blend of altruistic and imperial motives, played out through seeds." The legacy of the Green Revolution has been soundly deplored. Early on, Sir Otto Frankel referred to it as "a flood of vast dimensions." Even in the 1970s, reports of food shortages and famine began to surface, due to the failure of some of the new hybrid strains, and as reserve seed supplies were used for food, farmers had less seed available for planting. As a result of the new untested miracle varieties and the spread of monocropping (planting one variety), many developing countries experienced a tremendous loss of seed diversity. Whereas thousands of varieties of rice were once grown in the Philippines, today only two varieties account for 98 percent of the harvest. Of 10,000 varieties of wheat cultivated in China in

1949, only 1,000 remained 20 years later. In just a few decades, Mexico lost 80 percent of the varieties of maize once planted.

An essential characteristic of the new high-yield hybrids is that these varieties were developed to perform in tandem with heavy inputs of chemical fertilizers and pesticides, and abundant water. Debi Barker, in "Globalization and Industrial Agriculture," writes: "These mechanized, modern tools of farming—the seeds, the fertilizers, the chemicals, the farm equipment—quickly became commercial farming inputs that took control away from local farms and communities and gave more control to large corporate structures which supplied the inputs." Fowler and Mooney point out that the "fertilizer-seed partnership" was really the inevitable outcome of the export of corporate agricultural models to the developing countries. The standard argument, of course, was that greater yields would be the only realistic solution to feed more people (that argument is still made today). Ken Wilson notes: "Increasing yields in fields went side by side with decreasing resilience and sustainability in agricultural landscapes, and with a loss of cultural heritage." The post–WWII crop-breeding revolution, he says, was characterized by "the neglect of ten thousand years of indigenous farming traditions, and erosion of the benefits arising from successful co-evolution between landscapes, cultures, and food-ways." And, as Fowler and Mooney flatly state: "The green revolution failed to live up to its promise of solving the problem of world hunger."

Thirty years ago, Pat Roy Mooney observed that the

Green Revolution laid the groundwork for the global seed industry, an industry that has consistently ignored the cultural traditions and human rights of indigenous peoples, not to mention the diverse cultural expressions of plants. "If you control the seed . . . " it is obvious that you are in the best position to control the entire food system—regionally, nationally, or globally. The patient work of farmers, the practice of husbandry passed on through generations, the possibilities of creative exchange are superseded in favor of profit. According to Indian environmentalist and activist Vandana Shiva, "Without seed freedom there is no food freedom."

꒰

IN 1944, THREE scientists working at the Rockefeller Institute of Medical Research identified DNA (deoxyribonucleic acid) as a carrier of hereditary factors. Nine years later, James Watson and Francis Crick were able to describe the helical structure of DNA, a discovery that served as the basis for the transfer of genetic material from one organism to another. Twenty years after that, in 1973, two scientists working in California, Stanley Cohen and Herbert Boyer, successfully spliced a DNA sequence from an African clawed toad into bacterial DNA, and thus the era of genetic engineering was launched. The language used by the US patent office in awarding the patent—issued in 1980— aptly describes this new field of scientific inquiry:

> The ability of genes derived from totally different biological classes to replicate and be expressed in

a particular microorganism permits the attainment of interspecies genetic recombination. Thus, it becomes practical to introduce into a particular microorganism . . . functions which are indigenous to other classes of organisms.

Just to be clear, when speaking of plant breeding, "interspecies genetic recombination" is quite a departure from traditional breeding practices. Over the course of decades, transgenic engineers discovered a way to insert the DNA from one organism into that of another; natural processes do not allow such transfers of genetic material. The implications of this technology are not easily summarized, and once DNA, the "Philosopher's Stone of the new alchemists of modern industry" (Jack Kloppenburg's phrase), was decoded, the race was on to find lucrative applications. For the purposes of this chapter, we will have to be content to focus on the implications of the technology for the seed trade and for those who purchase and plant seeds—a vast subject in itself.

In 1980, in the case of Diamond v. Chakrabarty, the US Supreme Court, in a five to four vote, ruled that genetically engineered microorganisms could be patented. Now intellectual property rights extended not only to plants grown from seed but also to the genetic coding of basic life-forms. Almost immediately following the ruling, more than 1,800 patents were filed. Then, in 1992, the FDA, succumbing to pressure from George H. W. Bush's White House, announced that the biotechnology industry would not be burdened by new regulations. This determination, first put forward in the Coordinated

Framework for Regulation of Biotechnology in 1985, resulted in the principle of "substantial equivalence" between GE crops and crops grown using traditional breeding techniques. Since that time, the term "substantial equivalence," although debated with vehemence, has been repeatedly defended by the industry.

Shortly after the FDA's announcement, Calgene's Flavr Savr tomato, spliced with fish genes, became the first GE food crop to come on the market. Customers agreed on one critical fact—the tomato was tasteless—and this variety was soon abandoned. Flavr Savr was soon followed by GE canola and Monsanto's first "Roundup Ready" soybean, genetically modified to resist the application of Monsanto's foundational herbicide product, Roundup. Through genetic engineering, a foreign gene is inserted into the cells of the chosen plant, thus the seed will carry resistance to the herbicide. This is a radically different approach to weed management and indeed to crop husbandry as practiced for centuries. In 1993, Monsanto won approval for the introduction of recombinant bovine growth hormone (rBGH), which was injected into cows to increase milk production. In 1996, several varieties of corn that were engineered to contain Bt (*Bacillus thuringiensis*) toxins—meant to repel predator insects—were approved for release. Although the actual varieties of agricultural crops introduced by the biotech industry were limited—and still are—from 1996 to 2003, the worldwide acreage of GE crops rose from 2 million acres to 167 million acres. Worldwide, 90 percent of acreage planted in GE crops is concentrated in five countries—the United States, Canada, Argentina, Brazil, and China. As of 2007, one multinational US-based corporation, Monsanto, held the patents for 90 percent of all commercial plant traits.

Biotechnology, and specifically genetically engineered crops, was originally viewed by some as a magic bullet—the solution to feeding a growing global population. But in his book *The Botany of Desire*, Michael Pollan writes that "biotechnology has overthrown the old rules governing the relationship of nature and culture in a plant." Species have evolved separately, and over time, the physical structure of a species and how this organism relates to other organisms will change and adapt to other species evolving within a shared ecosystem. Through genetic engineering, the biotech industry is capable of introducing genetic material from one species into another—and a whole chain of interrelationships is significantly altered. Pollan notes that Charles Darwin was emphatic about one particular rule governing man's manipulation of nature: "Man does not actually produce variability."

"Now he does," Pollan says, observing: "The deliberate introduction into a plant of genes transported not only across species but across whole phyla means that the wall of that plant's essential identity—its irreducible wildness, you might say—has been breached, not by a virus, as sometimes happens in nature, but by humans wielding powerful tools."

Since its introduction, we have been led to believe that the science of genetic engineering is highly sophisticated and precise. Yet the truth is that the techniques used to transfer genetic material actually have a low rate of success; the results are often not predictable. Jack Kloppenburg quotes the University of

Wisconsin plant pathologist Robert Goodman, who "likens the random insertion of transgenes by genetic engineering techniques to throwing a grenade into the genome." The Harvard geneticist Richard Lewontin says that "the process of genetic engineering has a unique ability to produce deleterious effects and . . . this justifies the view that all varieties produced by recombinant DNA technology need to be specially scrutinized and tested." Ken Wilson, acknowledging that our debates about agriculture can easily become oversimplified, especially around "flashpoints like GMO crops," actually uses the word *clunky* to describe GE technology.[7]

Fred Kirschenmann, a North Dakota farmer and a long-time leader of the organic movement, exposes the impossibility of maintaining a segregation of biological organisms in nature once a new biological system has been introduced. He says: "In nature everything is connected. There are many pathways in nature, and the assumption that we can control all of the pathways is naïve in the extreme." Fedco Seeds founder C. R. Lawn, commenting on the nature of pollen, says: "You can't fence it in." This kind of informative testimony rises out of a counter-culture I have great respect for, though I am disappointed that the majority of American consumers simply do not hear it. Still, I am encouraged by the example, and the bravery, of Rachel Carson, whose individual voice effectively alerted a nation to the danger of our hubris. The final paragraph of *Silent Spring* deserves to be repeated: "The 'control of nature' is a

---

7   "This degree of intervention was hardly imaginable even a decade ago, and is far beyond the clunky genetically modified organisms (GMOs) that are currently so hotly contested." (From the foreword to *Where Our Food Comes From*.)

phrase conceived in arrogance, born of the Neanderthal age of biology and philosophy, when it was supposed that nature exists for the convenience of man."

In 2011, in a move that surprised many in the organic farming community, the USDA approved for release a number of new transgenic crops—alfalfa, sugar beets, and a corn variety intended for the production of ethanol. This followed a series of discussions that had been focused on developing a plan to encourage "coexistence" between GMO crops and non-GMO crops, assuming such a thing was possible. However admirable that goal may have been in theory, in reality, the genie had been released from the hopper. It is a known fact that once GMO canola plants find their way into non-GMO canola fields, the contamination will increase each year.[8] Just as worrisome is the tendency for some GMO crops to cross-pollinate with weeds of related plant families present in hedgerows or nearby fields (invigorating a new generation of "superweeds"). Years ago, before GE crops had been released into agricultural fields, Martin Alexander, a Cornell microbiologist, offered a word of caution, apparently unheeded by the industry:

It is difficult to see why a manmade genetic change would necessarily behave any differently from those

---

8   It is beyond the scope of this book to relate the story of the Canadian farmer Percy Schmeiser, but if the reader is not already acquainted with that story, please look into it, for it reveals a lot. I happened to have had a conversation with Mrs. Schmeiser outside of a lecture hall in San Rafael, California; she shared her husband's despair over the loss of their canola crop that was contaminated by genetic drift when a neighbor planted a GE variety sold by Monsanto. The loss of a variety nurtured for more than 40 years was devastating, though she was surprised and shocked that a multinational corporation chose to devote such considerable time and money to disrupt the life of a single farmer who had not purchased Monsanto seed.

occurring spontaneously in nature. . . . It is, thus, my view that alien organisms that are inadvertently or deliberately introduced into natural environments may survive, they may grow, they may find a susceptible host or other environment, and they may do harm.

In a plea for some sort of "meaningful coexistence plan" with the biotech industry, Charles Benbrook, chief scientist at the Organic Center, observes that "a deal with the devil was made in the early 1990s when the US government was an active accomplice in paving the way for near-total corporate control over seeds and plant genetic resources. That deal has led to profound injustices embedded in current US federal biotech policy." Given the green light, the industry has continued to seize control of the market.

Biotech companies have consistently portrayed themselves as friends of family farmers and the disenfranchised. The following quote from a former Monsanto CEO, Robert Shapiro, is a common campaign tactic: "As we stand on the edge of a new millennium, we dream of a tomorrow without hunger. To achieve that dream, we must welcome the science that promises hope. Biotechnology is one of tomorrow's tools today. Slowing its acceptance is a luxury our hungry world cannot afford."

Yet that acceptance has met with more resistance outside the borders of this country, notably in Europe. When a Monsanto advertising campaign was launched in England,

Prince Charles, a longtime supporter of organic agriculture, responded with an article titled "The Seeds of Disaster":

> I have always believed that agriculture should proceed in harmony with nature, recognizing that there are natural limits to our ambitions. . . . We simply do not know the long-term consequences for human health and the wider environment of releasing plants bred in this way. . . . I personally have no wish to eat anything produced by genetic modification, nor do I knowingly offer this sort of produce to my family or guests.

In pursuit of a greater share of the global seed trade, biotech corporations have repeatedly portrayed those critical of the technology as antiscience, irrational, or uninformed. Those opposed to the next miracle solution are criticized for lacking reliable scientific data to support their opinions. But in June 2012, a British organization known as Earth Open Source released a 123-page report titled "GMO Myths and Truths: an evidence-based examination of the claims made for the safety and efficacy of genetically modified crops." The report contains more than 600 citations and compiles data from physicians, scientists, government, and the media. One of the lead authors, Michael Antoniou, is the head of the Nuclear Biology Group at King's College London School of Medicine, with 28 years of experience in the use of GE technology;

another author, John Fagan, with a PhD in biochemistry and molecular and cell biology from Cornell, has been involved in the scientific debate over genetic engineering since 1994. I will quote from a list of "key points" taken from the much longer document:

> 1. Genetic engineering as used in crop development is not precise or predictable and has not been shown to be safe. The technique can result in the unexpected production of toxins or allergens in food that are unlikely to be spotted in current regulatory checks. . . .
>
>   6. The US FDA does not require mandatory safety testing of GM foods and crops, and does not even assess the safety of GM crops but only "deregulates" them, based on assurance from biotech companies that they are "substantially equivalent" to their non-GM counterparts. This is like claiming that a cow with BSE is substantially equivalent to a cow that does not have BSE and is thus safe to eat! Claims of substantial equivalence cannot be justified on scientific grounds. . . . [9]
>
>   10. Biotech companies have used patent

---

9   Michael Hansen, of the Consumers Union, points out: "The principle of substantial equivalence is an alibi with no scientific basis created out of thin air to prevent GMOs from being considered at least as food additives, and this enabled biotechnology companies to avoid the toxicological tests provided for in the Food, Drug, and Cosmetic Act to avoid labeling their products. That's why we say that American regulations of transgenic foods violate federal law."

claims and intellectual property protection laws to restrict access of independent researchers to GM crops for research purposes. As a result, limited research has been conducted on GM foods and crops by scientists who are independent of the GM industry. Scientists whose work has raised concerns about the safety of GMOs have been attacked and discredited in orchestrated campaigns by GM crop promoters.

11. Most GM crops (over 75%) are engineered to tolerate applications of herbicides. Where such GM crops have been adopted, they have led to massive increases in herbicide use.

12. Roundup, the herbicide that over 50% of all GM crops are engineered to tolerate, is not safe or benign as has been claimed but has been found to cause malformations (birth defects), reproductive problems, DNA damage, and cancer in test animals. Human epidemiological studies have found an association between Roundup exposure and miscarriage, birth defects, neurological development problems, DNA damage, and certain types of cancer. . . .

29. Conventional plant breeding, with the help of non-GM biotechnologies such as marker assisted selection, is a safer and more powerful method than GM to produce new crop varieties required to meet current and future needs of food production, especially in the face of rapid climate change.

And here is one dear to my heart:

> 30. Conventionally bred, locally adapted crops, used in combination with agro-ecological farming practices, offer a proven, sustainable approach to ensuring global food security.

An "evidence-based examination" is just what is needed right now, to counter the false and exaggerated claims of the biotech industry.

More than 100 US seed companies subscribe to the "Safe Seed Pledge," which eschews GMOs. Here is some of the language used in that pledge: "The mechanical transfer of genetic material outside of natural reproductive methods and between genera, families or kingdoms, poses great biological risks as well as economic, political and cultural threats. . . . More research and testing is necessary to further assess the potential risks of genetically engineered seeds."

It is, quite honestly, almost impossible to keep up with the volume of bulletins streaming through my mailbox. The opposition to the strong-arm techniques of the GE industry, and to the corporate dominance of the seed trade, is well informed and passionate, though lacking the deep pockets and the lobbying power of the corporations. The Center for Food Safety, Jeffrey Smith's Institute for Responsible Technology, and the Non-GMO Project (among others) are all focused on correcting the imbalance in the system and on challenging the safety and utility of GE crops and seeds. That theirs is an uphill battle is an understatement. When the GMO labeling bill, Proposition 37, was placed on the ballot in California in the autumn of 2012 (64 other nations in the world already have established labeling laws), the proposition was defeated, although polls showed that 90 percent of citizens support labeling. The biotech industry outspent supporters of labeling by five to one; Monsanto alone contributed $8.13 million (of a total of $46 million spent) to defeat the proposition.

Marie-Monique Robin, a French journalist and documentary filmmaker, quotes from a 2005 Pledge Report, in which Monsanto lists the values that they see as the foundation of their work: "integrity, transparency, dialogue, sharing, respect." In her important and troubling book, *The World According to Monsanto: Pollution, Corruption, and the Control of Our Food Supply*, Robin travels the world and the Internet in her search for signs of that integrity and respect, without much luck; instead, she has discovered just the opposite. Since the early 1990s, the Center for Food Safety (CFS) has tracked the Monsanto Corporation and issued a report in 2005 detailing the 144 lawsuits brought by the corporation against individual farmers (perhaps in the spirit of dialogue?). The lawsuits were filed against farmers who, in most cases, had never entered into a contractual agreement with Monsanto and who were unable to defend themselves in court against a powerful corporation. Patented seed found in any farmer's fields—usually the result of genetic drift, outside the control of any agriculturalist—seemingly is a justification for a corporation to turn to the courts.

Because many other investigations of "seed piracy" were settled out of court, and these cases involved a confidentiality clause, we can only guess at the actual number of farmers targeted by the corporation. If we return to the basics of an agricultural heritage passed on from one generation to another, the question arises as to why a company that values sharing and respect would employ 75 investigators, hire Pinkerton detectives, and spend $10 million to undermine individual farm families instead of assisting them. The reality is that in cases that involve patented seed purchased to be used for one growing season (the farmer is required to sign a binding contract with the seed patent owner), the only thing that matters, according to Joseph Mendelson of the CFS, "is that the gene was found in a field: Whatever the reason, the owner of the field is held liable." Robin quotes a Monsanto representative, in response to a farmer claiming that he never signed a contract, as saying: "We own you—we own anybody that buys our Roundup Ready products."

The time has come for the gentle farmer to "hurt the tune a little." In fact, to offer a short break from the inevitable failures of a global corporation insistent on gaining more and more control of the market, I will return to the words of my ancestors ("What Cathal Said," in a version by Tess Gallagher):

> *You can sing sweet*
> *and get the song sung*
> *but to get to the third dimension*
> *you have to sing it*
> *rough, hurt the tune a little. Put*

enough strength to it
that the notes slip. Then
something else happens. The song
gets large.

Before encountering Robin's book, I was unaware that Monsanto was founded by John Francis Queeny, a self-taught chemist, more than a hundred years ago, in 1901 (he named his company after his wife, Olga Mendez Monsanto). Shares in the company were sold on the New York Stock Exchange in 1929, and by the 1940s the company was a major producer/supplier of rubber, plastics, and synthetic fibers and had established a monopoly of the PCB (polychlorinated biphenyl) market. The company that now employs more than 17,500 people and operates facilities in 46 countries has, for much of its history, actively researched, manufactured, and marketed chemical pollutants that have since been banned. Monsanto produced PCBs (polychlorinated biphenyls) for decades until they were banned in the early 1980s; 2,4,5-T, an herbicide containing dioxin, the basis for Agent Orange, which was finally banned in the mid-1970s; and DDT, which was banned in the early 1970s largely due to the investigative journalism of Rachel Carson. Marie-Monique Robin follows the path of the company through the manufacture of these chemicals, leading up to the present-day production and distribution of Roundup. Each step on the pathway deserves a chapter (Robin supplies it) to recognize the manipulation and deception required to market a product on a vast scale without a full understanding of the long-term effects of such a commercial action. Agent Orange—

used as a defoliant, along with other dioxin products, on 8 million acres in Vietnam between 1962 and 1971, an estimated 20 million gallons—was put in use to win a war that was lost. (The significant danger that dioxin use posed for human health was not publicly acknowledged until the US secretary of agriculture called for the suspension of the use of the product known as 2,4,5-T in 1970.) Roundup, the trade name for the herbicide glyphosate, discovered in the late 1960s by Monsanto chemists, first applied to soil in liquid form and now genetically engineered into the plant, is the method of choice in the war against weeds. Brian Tokar, in an article titled "Agribusiness, Biotechnology, and War" in Z Magazine, finds the connection chilling: "Virtually all of the leading companies that brought us chemical fertilizers and pesticides made their greatest fortunes during wartime." Monsanto, one of these companies, has now turned its attention toward seeds. In her investigation of the company, Robin uncovers an all-too-familiar pattern of obfuscation, false advertising, and concealment; lawsuits and court battles have become an integral part of the conduct of the business.

The question now arises: Is it at all wise or beneficial for a corporation with a scarred legacy of creating chemical solutions for complex problems to control almost one-third of the global seed trade? How will the millions of indigenous farmers, gardeners, breeders, and creative horticulturalists find freedom and space to experiment and to promote diversity? We need more of it, not less. In the words of Yudhvir Singh, a spokesman for the north Indian Bharatiya Kisan Union (as quoted by

Robin), it is an extreme challenge to confront a multinational corporation that is skilled at creating its own advantage, with the intent to "seize control of the seeds and hence of the food of the world."

In the spring of 2013, the latest in a long series of alerts concerned with the behavior of this corporation appeared as an earmark in the Senate fiscal year 2013 Agriculture Appropriations bill, part of the Continuing Resolution (an unusual congressional compromise agreement—a temporary fix—keeping the US government in business). Inserted into the bill with little or no committee discussion, and with various lawmakers claiming ignorance, the "Monsanto Rider," or the "Monsanto Protection Act," directs the secretary of agriculture to grant a temporary permit for a GE crop to be purchased, planted, and cultivated, superseding any verdict a federal court might impose. The Center for Food Safety correctly labels this action as "corporate welfare," though the rider is actually named the "Farmer Assurance Provision." Perhaps farmers who have signed the Monsanto Technology/Stewardship Agreement (MTSA) should have been consulted about the wording. In order to purchase GE seeds, a farmer must sign a MTSA agreement that prohibits him or her from replanting (so that the farmer must repurchase seed each year), limits the corporation's liability for the product, requires that only Roundup be used as an herbicide, and gives the corporation's agents access to the purchaser's fields. The ETC Group (Action Group on Erosion, Technology, and Concentration), a civil society organization, has used the term "bio-serfdom" to describe the relationship.

In a paper presented at the 1998 American Seed Trade Association annual meeting, Cornell plant breeder Ronnie Coffman offered some prescient comments, using a metaphor that carries particular resonance for me:

> If words were copyrighted, only the few who owned them could communicate and our society would be harmed. Genes are analogous to words in that they allow the creation of new plant cultivars just as words allow the creation of a book. Everyone in society should have the right to use genes. Cultivars (novel genotypes or combinations of genes), not genes, should be eligible for patenting. It is now clear that the patenting of genes will result in only two or three companies having a major influence on the food system.

❧

Writing about the future of agriculture in 2004, which he believes is "not only the oldest but also the most important of humanity's productive activities," Prince Charles observes that "manipulating nature is, at best, an uncertain business." Environmental journalist Richard Manning writes: "If there was a key mistake of the green revolution, it was simplifying a system that is by its very nature complex." I am hardly objective on this topic, but increasingly it appears that an agroecological[10] approach to managing food systems and seed supplies is not only a practical solution but an affordable one. Charles Benbrook advises:

> Without a doubt most of the progress that's going to be made in promoting food security in many parts of the world will be grounded in agroecological methods. These methods are best for increasing soil fertility, combating climate change, and assuring that farmers have a variety of crops that survive in drought and wet years. The majority of progress in areas dealing with recurrent periods of severe food insecurity won't depend on resources and technologies from outside the region, including GM seeds.

I see nothing wrong with acknowledging the complexity of natural systems, nor of recognizing that in order to encourage diversity we must respect that complexity and take a whole-systems approach. The one pest–one chemical model did not work, and the one pest–one gene approach does not look promising either (despite the propaganda). Fred Kirschenmann believes that we have three choices open to us—to stay the course and continue our method of "resource-exploitation agriculture," to opt for the next "technological fix . . . transgenic technologies," or to "return to some model of agricultural restraint" that would require a more complete

---

10   Agroecology is a whole-systems approach to the practice of agriculture; the farmer takes into account the entire ecosystem when making decisions about tillage, crop rotation, cultivation, irrigation, and seed selection.

understanding of the integrity and fluidity of natural systems. I favor the latter approach—and if we surveyed the many agriculturalists encountered in this chapter, they would no doubt be in complete agreement.

In 2011, a preemptive lawsuit was filed on behalf of 83 plaintiffs in response to the strong-arm legal tactics repeatedly employed by Monsanto to intimidate farmers. In the case Organic Seed Growers & Trade Association et al. v. Monsanto, the plaintiffs, represented by the Public Patent Foundation, a nonprofit that specializes in patent law cases that involve the public interest, have sought a declaratory judgment that would prevent the biotech corporation from suing the plaintiffs for patent violation. At issue for organic farmers, and for the organic community, is the threat of contamination by GMO seeds planted in neighboring farm fields. In February 2012, Judge Naomi Buchwald, in a hearing held in New York City, ruled in favor of the industry giant, dismissing the case. She determined that the plaintiffs had no standing to sue, and referred to the suit as "a transparent effort to create a controversy where none exists." In June 2013, the court of appeals for the federal circuit in Washington, DC, ruled that the lawsuit would not be allowed to continue, citing a promise made by the biotech corporation: "Monsanto has made binding assurances that it will not 'take legal action against growers whose crops might inadvertently contain traces of Monsanto genes (because, for example, some transgenic seed or pollen blew onto the grower's land).'" Given how vague the promise is and Monsanto's previous record, it is hard to feel assured of its enforceability.

More than 270,000 people were represented in this class action lawsuit on the side of the OSGATA. Isaura Andaluz, speaking for plaintiff Cuatro Puertas, said: "Seeds are the memory of life . . . if planted and saved annually, cross pollination ensures the seeds continue to adapt." To protect diversity in our food crops, we will have to maintain a broad genetic base of plants. Frank Morton, of Wild Garden Seed, also a plaintiff in this case, emphasizes, "the more narrow the genetic material, the more dangerous it is."

At the close of this long chapter, I would like to return to the seed source we began with, Fedco of Maine. When the wholesale seed supplier Seminis was purchased by Monsanto in 2010, Fedco made the difficult choice to boycott the company, despite the fact that 70 Seminis varieties listed in the catalog were generating 11 percent of the co-op's revenue. C. R. Lawn explains Fedco's choice: "For make no mistake. There can be no sustainable food system or sustainable seed system as long as our farms are dependent on multinationals like Monsanto and Syngenta." Noting the significant revival of interest in organic seed, heirloom plant varieties, and ecological growing methods, C. R. discovers some real promise: " . . . this is the most fragile of networks. It is being built up skill by skill, variety by variety, grower by grower, like a good foundation, one brick at a time." Bricks are made of clay—a soil substance produced over time by the interaction of earth forces. The memory of life held within a seed is as fragile as this action, but planted and tended with care, and freely exchanged, the memory will endure.

*Health is the capacity of the land for self-renewal. Conservation is
our effort to understand and preserve this capacity.*

—Aldo Leopold, *A Sand County Almanac*

CHAPTER 8

Svalbard

DURING THE GENTLE SPRING OF 2008, I was invited by my
daughter's science teacher to speak to her high school class
about my experience working with plants and seeds. Many of
the students had visited our community farm to assist with
seeding and transplanting, so we had a platform—a shared
seedbed—for our discussion. My intention was to offer an
overview of the kind of agriculture we practice and to discuss
how we select and seed plant varieties to maintain diversity in
our fields. I also wanted to share the news of a significant event
I supposed they would be unaware of—the opening of the first
global seed bank in Svalbard, Norway. Though it would be

difficult to reach 12th grade without some study of the structure of flowers and the production of seeds, I figured that no student had yet heard of a "seed bank" or contemplated the need or function of such a thing. Yet from my first mention of John Seabrook's instructive story "Sowing for Apocalypse," which appeared in the *New Yorker's* Annals of Agriculture, I had—to the delight of one father and educator—captured their complete and rapt (teenage) attention.

Five hundred miles north of mainland Norway, above the Arctic Circle, in a group of islands known as Svalbard, a most unusual event had occurred just weeks earlier: The Global Crop Diversity Trust, directed by Cary Fowler, opened the first designated global seed bank on Spitsbergen, near the town of Longyearbyen. In the world's northernmost town, the midnight sun lasts from April 20 until the 23rd of August, and the polar night endures from October 26 until February 15. The annual average temperature in Svalbard is 25 degrees (-4 C), and the record low is -51 degrees (-46.3 C). I have not yet visited, but my friend and Amagansett neighbor, the ocean conservationist author Carl Safina, has; I admire his writing, so to set the scene I will quote from his recent book, *The View from Lazy Point*:

> The islands collectively called Svalbard rise from the sea so far north they seem to lie beyond human thought. . . . The nearest tree—or even shrub—is hundreds of miles behind us, on the European continent. . . .

Through what looks like a concrete-reinforced cellar door in a frozen mountain, we enter a nearly four-hundred-foot-long tunnel dug through solid rock. The tunnel connects a series of cavernous storage rooms maintained at 0 degrees Fahrenheit (-18 degrees C), and the whole space resembles a big multichambered cave. This is the Doomsday Vault, the world's cold-storage site for agriculture, a safety-deposit box for seeds.

At present, this underground vault has the capacity to store almost two billion seeds, though Cary Fowler is not fond of the popular "Doomsday" designation, which he feels wrongly emphasizes a bleak outlook rather than a carefully laid plan to support global food security for a world that still must consider harvests. To Fowler, who spent more than 25 years preparing for this event, the seed bank is a calculated attempt to save the biological foundation of agriculture: crop diversity.

Though Nikolai Vavilov is credited with being the first to dream of a true world seed bank—in the early 20th century, the Vavilov Institute of Plant Industry (named the All-Union Institute of Applied Botany and New Crops at the time) was the most respected seed bank in the world—it took decades of determined effort on the part of Cary Fowler and many others to realize Vavilov's dream. Fowler, who grew up in the American South with a love of agriculture, worked as a researcher in the mid-1970s for Frances Moore Lappé on a book that she coauthored with Joseph Collins, *Food First*; he then teamed up with the Canadian activist Pat Roy Mooney to focus on international food policy issues. Early on, Fowler and Mooney sought to expose the complicated problems arising from the patenting of seeds. With the passing of the US Plant Variety Protection Act (PVPA) in 1970, which gave patent protection to those who developed novel seed plant varieties, the ownership of genetic resources, and thus of "seed rights," entered an entirely new era.[1] Mooney published *Seeds of the Earth: Public or Private Resource* in 1979, and he coauthored the masterful *Shattering* with Fowler in 1990, which laid out their vision to reverse the trend that could lead us toward "the shattering of agriculture itself."[2] The warning they issue is a stark one: "What is at stake is the integrity, future and control of this first link in the food chain."

Since the early 20th century, a select few botanists and plant collectors had been raising concerns about genetic erosion. One of these was Erna Bennett, a plant breeder from Ulster, Ireland, who worked for the FAO (Food and Agriculture Organization of the United Nations) collecting and selecting seeds in Turkey, in a region that Vavilov identified as a center of diversity. In 1970, Bennett alerted the scientific community to the growing crisis of genetic erosion when she

1   In 1980, the act was extended to include more species, and in the case of Diamond v. Chakrabarty referred to earlier, the court decision opened the way for the patenting of forms of life based on their genetic coding.

2   *Shattering* refers to the natural process of a seed head opening up and scattering its seed. Early domestication of grain plants involved saving seed that tended to adhere to the stalk.

coedited, with Sir Otto Frankel, the first classic text on genetic resources. They were following the work of Vavilov, who had been cognizant that genetic erosion was a real threat to global food security. Another veteran plant explorer, Harry Harlan, a supporter and ally of Vavilov, had issued a warning in 1936 that the introduction of new crop varieties, developed for monocultural use, was accelerating the loss of our genetic diversity. Few were listening, however, and public and private funding were increasingly directed toward the development of high-yielding crops while "genetic conservation"—a term coined by Bennett—was neglected.

As the Green Revolution began to sweep the globe in the 1970s, only a handful of gene banks and relatively few members of the scientific community were focused on the growing need to protect biological diversity. Beginning in that decade, according to Fowler and Mooney: "New advances in microbiology and new corporate opportunities in the seed industry, combined with the ecology movement that built up after the sixties, made germplasm conservation a social, political, economic, and corporate necessity." Scientific and technological innovation led to an increase in the development of hybrid seeds—and hybrids, derived by crossing inbred lines, have a narrow spectrum of genes. More and more traditional varieties were becoming extinct. Eventually, this led to a broader recognition that germplasm had to be protected to ensure that resources for new varieties would continue to be available.

In 1958, the United States opened the National Seed

Storage Laboratory (NSSL) in Fort Collins, Colorado. By the 1970s, this gene bank had become a depository for the world's major crops. The NSSL came to serve the same purpose on the world stage as had the Vavilov Institute in the 1920s. Ultimately, however, the NSSL, like any other publicly funded seed bank, is reliant on the political will of changing administrations to support seed conservation. Underfunded for years, the NSSL, part of the National Plant Germplasm System (NPGS), and operated by the USDA, was repeatedly labeled as inadequate in the 1980s. In 1992, Congress funded an addition to the facility, so that the modernized vaults can now store up to 1.5 million seed samples. Several years ago, the NSSL became the first seed bank in the world to utilize a storage technology known as cryopreservation—some seeds are stored (at -196°C) in steel tanks containing liquid nitrogen.

At the 1981 annual conference in Rome of the UN Food and Agriculture Organization, with encouragement from Fowler and Mooney, the Mexican delegation proposed the creation of a world seed bank, which eventually underpinned a document titled the "International Undertaking on Plant Genetic Resources for Food and Agriculture." The ongoing discussions, intense and fueled by conflicting opinions, centered on a very basic question, with nothing less than the ultimate control of the global seed trade in the balance: Who owns plant germplasm—the country of origin, or the nation with more advanced technology and money?

For their parts, Mooney and Fowler—along with much

of the conservationist community—view plants and seeds as part of the common heritage of the human race. For 10,000 years or more, indigenous farmers have been selecting and developing plants as a means of survival (and botanical curiosity). Our ancestors successfully domesticated and brought under cultivation about 200 species of plants long before modern plant breeding existed. (Remember Darwin's description: "cultivating the best known variety, sowing its seeds, and, when a slightly better variety chanced to appear, selecting it, and so onwards.") Of the 20 crops that supply most of our food, all were developed by traditional farmers—by saving seeds that feed us now—who had never given a thought to plant breeders' rights (PBR).

By the mid-20th century, the more powerful nations of the North had recognized that control of plant germplasm in developing countries was critical to establishing dominance within the seed trade. Corporate seed interests tended to view "common heritage" as an affront to intellectual property rights—the recognition, in legal terms, of the investment of time and money on the part of these international seed companies. Representatives of third world countries at the FAO supported the idea of a genetic commons—seeds should always be considered part of our collective heritage. Fowler and Mooney shared this view: After centuries of a free flow of genetic material from the South to the North, it was absurd that the Northern seed trade would now claim ownership. Jack Ralph Kloppenburg, writing in 1988, assessed a 20-year trend in the geopolitical climate:

Third World nations found their own genetic resources, albeit transformed by plant breeders, confronting them as commodities. This pattern has been seen as doubly inequitable because the commercial varieties purveyed by the seed trade have been developed out of germplasm initially obtained free from the Third World.

Developing countries reasoned that if the Northern countries insisted on claiming ownership of patented seeds, they could assert ownership of the genetic material collected in the South. To bolster the mission of an international seed bank, the Mexican delegation at the FAO proposed free access for both North and South to the seeds to be held within that depository—patented or not.

By 1984, the dispute had grown so bitter that the *Wall Street Journal* dubbed the controversy over the public/private ownership the "Seed Wars." Cary Fowler claims that when he first appeared as a public advocate at the FAO meetings, he was labeled a "dangerous radical." As a child of the sixties, inspired by poetry and "the perennial philosophy," I am all too familiar with this moniker—and know well how perfectly sensible concepts that prove to have social and historical validity in the improvement of our human condition often appear to be dangerous or radical to the ruling order.

A few years later, in 1993, the FAO hired Fowler to oversee the drafting of a global plan of action; the end result would be a document—adopted by the United Nations in 2001—that contained the beginning legal framework for a world seed bank. The finished document was something of a compromise—farmers' rights had survived, ensuring that farmers would receive some compensation for genetic material collected by the Northern seed trade, although the developing countries agreed not to press their opposition to seed patenting.

Norway had proposed building a world seed vault at Svalbard when the skirmishes of the seed wars were raging, but it wasn't until the UN treaty was ratified that action was possible. A Nordic Gene Bank already existed there, in an abandoned coal mine, which served as a repository for seeds from five countries—Norway, Denmark, Finland, Iceland, and Sweden. Norway offered seed money and support in an effort to provide stability to the 1,700-plus gene banks scattered around the globe in more than 100 countries. Cary Fowler estimates that half of these seed banks—often created for short-term storage—exist in somewhat precarious conditions, due to inadequate funding, human conflict, or lack of government support.[3] At Svalbard, a major portion of the estimated 7.4 million seed samples stored in a multitude of collections can eventually be conserved deep in a permafrost mountain.

---

3   One notable exception is the Millennium Seed Bank Project, managed by the Royal Botanic Gardens, Kew—an international conservation initiative with 16 participating countries—which opened in 2000. Millennium's focus is on the conservation of the seeds of wild plant species; by 2010, 24,200 species had been conserved, or 10 percent of the estimated species of seed plants in the world. The UK Flora Programme of the MSBP has collected seeds from 96 percent of the United Kingdom's plants, about which, Wolfgang Stuppy observes: "This is the first time that any country has underpinned the conservation of its flora in this way." The £72 million cost of the project has been borne by the Millennium Commission and a consortium of corporate and private sponsors.

The Norwegian Ministry of Agriculture and Food holds the legal and financial responsibility for the seed vault, and the day-to-day operation is jointly managed by the Norwegian Ministry, the Global Crop Diversity Trust, and the Nordic Genetic Resource Center. Designed to eventually store up to 4.5 million varieties—nearly two billion seeds—by February 2012, only 4 years after opening, the Svalbard bank housed 740,000 seed samples (among which can be found, for example, 40,000 varieties of beans and 140,000 varieties of wheat). "A library of life . . . " is the phrase Cary Fowler chooses to describe the genetic material lodged on shelves within the frozen earth adjacent to the world's northernmost town.

To reach the Svalbard vault, seeds destined for storage inevitably travel great distances, as they have done for centuries, though previously in a less-organized fashion. (I have to pause here to again acknowledge the ingenuity of the angiosperm revolution that gave us the packaged and preserved energy of plants.) As yet, not a single box of seeds has been lost in transport. Fowler calls it "a miracle of coordination." Conserved in the library of life, the story inherent in each seed has been brought to safety, for a time. Seeds are stored in sealed glass containers, inside sealed boxes, inside an air-lock chamber, behind locked doors. How useful or viable all of this genetic information will be for a future climate is an unanswered question. "Seeds are placed in gene banks not so much to preserve seeds as to preserve diversity," Fowler and Mooney noted in *Shattering.* "Because extinction is forever, conservation must be forever."

✣

THIS GRAND ACHIEVEMENT is not without its critics. F. William Engdahl, author of *Seeds of Destruction: The Hidden Agenda of Genetic Manipulation*, questions the intent of the Gates and Rockefeller foundations—who have backed the proliferation of patented seeds—in donating millions of dollars to support the project. Andrew Kimbrell, executive director of the Center for Food Safety, is also troubled by million-dollar donations from biotech developers Dupont and Syngenta. He worries that the Svalbard seed deposit is "one stop shopping for the corporations," that biotech firms will now have easier access to a wide range of genetic material.

However, Cary Fowler insists that corporate donations represent less than 2 percent of the total funding for the Global Crop Diversity Trust and that donations come with no strings attached. Most of the seeds stored in the Svalbard vault are already part of other publicly available collections, and because of their "black box" policy, only the original depositor has direct access to their seed material (breeders or researchers must request samples from the country of origin). Besides, Fowler continues, it is unlikely the Norwegian government would facilitate an easy exchange of genetic material to benefit biotech interests, as GE crops are not even allowed in that country.

For Kimbrell, however, the problems associated with Svalbard outweight the advantages: "You want to protect diversity in the habitat where it lives."

The debate will continue over which method of biodiversity

conservation is more effective—in situ *or* ex situ—but Gary Paul Nabhan advises that these two "strategies must be used in a complementary manner." Seed saving is an ancient art, but as stewards of the remaining biodiversity on earth, it is incumbent upon us to recognize the strengths and weaknesses of both conservation approaches: "If we wish to conserve useful plant species for posterity, we will do well to learn the risks that are involved with every available strategy, none of which are foolproof or flawless," he says.

In February 2013, the Seed Savers Exchange, an organization with enough experience to operate its own seed bank, sent a sixth shipment of seeds to the Svalbard vault—366 varieties in all, bringing their total deposit in the collection to 2,248 varieties. SSE acknowledges the Svalbard bank as "the ultimate in long term storage for seed," and because of the black box agreement, the SSE is confident that they will have free access to the seeds they have donated.[4]

The creation of a world seed bank was something that Vavilov had envisioned almost 100 years ago. Cary Fowler has spent more than 25 years realizing that dream, and intense debate has followed the vision from the beginning. In an interview published in the *Atlantic*, 4 years after Svalbard's opening, Fowler said:

> We're not saying that we have a crystal ball and that
> we know what's going to happen and we know what's

---

4   For the reader who desires to investigate further, search for Kent Whealy's commentary on the connection between SSE and the Svalbard seed bank. That level of internal politics is beyond the scope of this book.

needed, but we do know that the diversity we have represents an immense number of untapped options, and what we're trying to do is keep all of those options. I think it was Paul Ehrlich who said "the first rule of intelligent tinkering is to save all the parts," and that's what we're trying to do.[5]

In terms of biological and cultural conservation, the success of the Svalbard experiment will depend upon how closely the project adheres to the indigenous roots of plant husbandry and seed preservation. Emigdio Ballon, keeper of the seeds at the newly constructed seed bank Tesuque Pueblo in New Mexico, expresses the essence of what a grower or a seedsman desires to pass on to his children: "If you have seeds you can survive. . . . Everybody can have access to this seed bank." Holding seeds in his hands, Ballon says: "They have inside the seeds almost all the expression of the land."

At the Svalbard seed bank, we must hope that the most precious natural resource we have will be well cared for. There is such beauty in the simplicity of a single seed—which, again, is really the embryo of a plant, packaged energy surrounded by a hardened and waterproof shell. Some embryos can last a very long time, but not forever. Our first seed saving ancestors, though aware of the invaluable currency of plants, did not need to envision a strategy for long-term conservation. We do—in order to ensure plant diversity and a supply of food for the coming generations.

---

5   It was actually Aldo Leopold who wrote that "to keep every cog and wheel is the first precaution of intelligent tinkering."

"What I say is that, if a fellow really likes potatoes,
he must be a pretty decent sort of fellow."

—A. A. Milne, *Winnie-the-Pooh*

CHAPTER 9

"A Kind of Earth Nut"

RETURNING FROM THE FIELD after a June seeding of a second round of bush beans—'Jade', 'Purple Burgundy', and 'Golden Rocky'—and 'Over the Rainbow' carrots, cosmos flowers, and 'Autumn Beauty' sunflowers, I open the small leather-bound notebook that holds a reserved spot on the mantel of the rolltop desk I inherited from my father, artfully manufactured by the Indianapolis Cabinet Company. Several years ago, I carried this notebook to the banks of the Wisconsin river, to the place that inspired Aldo Leopold to compose his "Land Ethic." A half dozen pine quills still lodge in the spine to illustrate questions raised upon a visit to Leopold's shack: What does it mean to work the land, to conserve land, and to recognize the soil as a fountain of energy?

I consider myself fortunate that these questions animate my work, or at least my mind, as I go about the tasks of seeding and cultivating. I am stimulated to practice good husbandry—to nourish the soil, the seedbed for plants that will bear fruit and nourish us in return. This aspect of the job is timeless; I am linked with planters and cultivators throughout a millennium or more.

In the next few chapters, I hope to communicate the experience I have gained *on the ground* as a steward and a seedsman, a participant in the cycle of the seasons, a witness to seeds at rest, at germination, and in formation as the expression of a plant that evolves to "go to seed." In the words of the philosopher Alfred North Whitehead: "There is no substitute for the direct perception of the concrete achievement of a thing in its actuality."

PRONOUNCE THE WORD *seed* and the image it conjures, for most people, is that of a round pebblelike thing that can be held between one's thumb and index finger. It is not likely to call to mind the hefty, hard shell of a coconut palm or the seed of the sea-bean that is carried in a 4-foot-long pod. Nor is it likely to evoke a piece, or a cutting, of a potato. While it is true that the common potato, *Solanum tuberosum*, is primarily reproduced vegetatively (via a cutting of the tuber), when we engage in grading, cutting, planting, and cultivating potatoes, we always speak of the tubers we handle as "seed." And when I have my annual conversation with Allison, the Maine Potato Lady, we are most assuredly discussing seed potatoes, not those we will later serve up for the family dinner.

Having been introduced to the art of planting potatoes in Cornwall, England, where the trick was to encourage eyes to sprout on cuttings (an egg-sized potato is ideal for seed—larger potatoes are usually cut weeks before planting, so that the eyes will naturally sprout), I was slow to discover that this member of the Solanaceae family can indeed reproduce and multiply via seeds rather than cuttings. Occasionally, the varieties that we plant in Amagansett do produce fruit and seed, though modestly compared to their wild relatives and most early domesticated varieties, which produce cherry tomato–like fruits that contain up to 200 seeds. Suzanne Ashworth, author of *Seed to Seed*, the invaluable book for seed savers, calls these fruits "seed balls." The home gardener or farmer will most likely always grow potatoes from tuber cuttings (as I was taught, and continue to do at the farm, where we plant long-domesticated varieties), thus the plant that matures is a clone of the parent plant. When planted from the actual seed—saved from the seed ball—the result is unpredictable, not typically a favorable scenario for a gardener. Each seed will produce a different variety of potato, which may be precisely what a breeder desires, but for standard production such an outcome would be best described as confusing.

By the first of November, we are often still harvesting

potatoes from the field we have named Birch Hill. With thousands of pounds of spuds stored in the ground (though the plants die back months earlier, the tubers keep well under the soil), we wait for the window for harvesting, which hinges on the weather. An element of risk is involved in our methodology, as in all other aspects of the farming profession—it is indigenous to the task. A light frost could soon arrive, even here in Zone 7, but we are relatively safe unless Jack Frost decides to take up residence for an entire week. We are the proud owners of a traditional, let us call it an heirloom, root cellar—in other words, the temperature is not regulated. Storing the tubers in the ground has worked well for years; and so we are obligated to delay our harvest until the root cellar has reached a proper temperature for winter storage (around 40 degrees).

So when November arrives, almost 6 months after planting time, we are ready to uncover the story of each storage seed cultivar. In April, we plant 20 or so varieties, and from July on we inspect and harvest for market with a digging fork. Yet not until the mechanical digger lifts the tubers from the long rows will we know the yield of our storage varieties. Early in the harvest season, our community farm members also use forks to find the potatoes first to mature: 'Red Gold', 'Chieftain', 'Yellow Finn', and the fingerlings—'Laratte' and 'French Red'. Over the years, we have learned which varieties will keep well in the earth and will store in the cellar until April (a year after planting): 'Kennebec', 'Keuka Gold', 'Carola', and 'Bintze'. 'Bintze', a Dutch variety dating from the late 19th century, has the advantage of being the least favored by our insect nemesis, the Colorado potato beetle (CPB). The CPBs seem to adore the tall, elegant plants produced by 'French Red' fingerlings, but they will ignore the less frivolous leafage of the Dutch cultivar. In our ongoing showdown with this pesky beetle, we seek every available competitive advantage.

I have written elsewhere about this "propitious esculent" (a term I borrow from the subtitle of John Reader's marvelous book, *Potato*), and here I would like to correct one small detail. For many years, I was under the impression—gained through observation and conversation—that the CPB was only able to reach a destination on foot, or to be precise, on six legs. If this was not true, why would Cornell Cooperative Extension introduce the "trench method" to discourage the onslaught of the beetles?[1]

Then, several springs ago, I received a call from Kip, a Cornell graduate student conducting research on CPBs. She contacted me just after we had planted seed potatoes in April, well before the first sprouts had emerged. "I hope it is okay for me to place a few trap plants here and there to catch the first fliers?" she asked.

---

1   This method is worth a footnote, as a rather comic, if well-intentioned, measure to control an insect adversary. A trench is cut, by machine, along the edge of a planting field, as a barrier between the field and the hedgerow. The trench is lined with black plastic, a slippery substance for a beetle. It is well known that CPBs overwinter in hedgerows; when they emerge in spring they march, in infantry fashion, toward the emerging solanaceous plants. When their toes touch the plastic they slide into the black valley; even the most athletic of beetles is unable to climb up and out (at least in theory). If you are fearful at the notion of hordes of insects usurping your domain, I would not recommend a viewing of this spectacle—thousands upon thousands of beetles attempt to overtop one another, only to tumble back into the heart of the trench.

"Fliers?" I had to respond. "But I thought they were only able to walk?"

"Oh no!" Kip exclaimed. "Though many beetles are walkers, the advance guard are very adept at flying!"

After a moment's pause, a scorecard flashed before me—advantage, beetles.

❦

AS APPEALING AS the potato is to these beetles I have come to know well, this esculent (an edible plant) is equally appealing to the human family. This has not always been the case, though our relationship with this prolific food plant spans centuries and includes nearly every country of the world. I am keen on the subject of potatoes, not simply because my mother, Mary Pratt, was born in Ireland, but also because of the role played by this one plant in the fields that feed so many of us. In *Shattering*, Cary Fowler and Pat Roy Mooney highlight its significance:

> Potatoes were the first crop in modern history to be devastated by lack of resistance—and the first crop to be rescued by the wealth of defenses built up over thousands of years in its center of diversity. Thus the Irish potato famine stands both as the most dramatic warning of the dangers of genetic uniformity and the clearest example of the value of preserving diversity.

It is commonly accepted that *Solanum tuberosum* originated in the Andes mountains about 10,000 years ago, though widespread domestication probably was not established until several thousand years later. Tracing the wild ancestor of any plant requires extensive detective labor and also some "inspired guesswork," in the words of J. G. Hawkes. Professor Hawkes, a professor of botany who wrote his PhD thesis at Cambridge on the taxonomy of *S. tuberosum*, identified 169 species of wild potato, and these relatives of the potato we now know exist over a vast geographical area. The most primitive relative is a small air plant (known as an epiphyte) named *S. morelliforme*, which grows on mossy branches of oak trees in southern Mexico. Many of the ancestors of our potato—technically *S. tuberosum* subsp. *tuberosum*—contain toxic glycoalkaloid compounds, so the Andean farmers who originally bred this tuber had to possess significant patience. By the time the Incas came to dominate the Andes in the late 15th century, the potato was established as a primary food crop. Although maize was the dominant crop of the lowlands, the versatile potato was the chosen cultivar of the highlands. The Incas became very sophisticated farmers, and the legacy of their experimentation is relevant today—5,000 or so varieties of potatoes remain in cultivation. Through continuous trials in the high fields of the Andes, these savvy farmers created resilient cultivars.

It is not a simple task to trace the origin of this plant. The early commentary can easily be confused with references to the sweet potato, *Ipomea batatas*, which is properly a member of the Convolvulaceae family. The first mention of the potato by name occurs in the writings of one Pedro Cieza de León, a Spaniard who traveled widely in the Andes in the 1530s and 1540s. Noting the principal food of the indigenous people, he writes: "One is called potato, and is a kind of earth nut, which, after it has been boiled, is as tender as a cooked chestnut, but has no more skin than a truffle. . . ."

The returning Spanish explorers/conquerors introduced the newfound tuber to Spain sometime around 1570, though this oddity from the New World remained a kind of garden curiosity in the Old World until the mid-17th century—at which point, the versatile food from the mountains of South America found a natural home on Europe's Emerald Island.

The potato first appeared in Ireland in the late 1580s, and within the course of several decades it had become the prime food source for the common Irishman. Throughout Europe, wherever an ample supply of grain for bread existed, people were not overly interested in growing potatoes. But with the failure of grain crops, the interruption of repetitive wars, and the inequity of land distribution, the benefits of *S. tuberosum* as a food crop became increasingly apparent. Eventually, as William H. McNeill, a historian who published a paper in the journal *Social Research* in 1999 titled "How the Potato Changed the World's History," writes, "The European scramble for empire overseas, immigration to the United States and elsewhere, and all the other leading characteristics of the two centuries between 1750 and 1950 were fundamentally affected by the way potatoes expanded northern Europe's food supply." It is a fairly predictable result of the global distribution of seeds and food crops that a tuber native to the highlands of South America could find such success within a relatively short span of time, but it is also remarkable. Perhaps this is what led Michael Pollan in *The Botany of Desire* to associate the potato with the desire he labels "control."[2]

In Ireland, of course, due to a complexity of socioeconomic conditions, a reliance on the potato as the major food source resulted in a disaster of historic proportion. The potato found its way to Europe before crossing the temperate zones that exist between South and North America, and it was especially at home in the cool climate of Ireland. Because the majority of the Irish people had to make do on marginal land that made cultivating grains difficult, this foodstuff arrived as a godsend. With the addition of cow's milk, which supplied vitamin A, potatoes could provide adequate nutrition—in the form of carbohydrates, protein, and vitamins B and C—to feed a growing population.[3] It was possible for the potato to yield 4 to 10 times as much food as an acre of grain. Unfortunately, those who tilled the soil and spread the seed did not do so with plant diversity in mind. The crop responsible for such a dramatic rise in population all derived from two parental lines; those who cultivated it at the time were ignorant of the danger. In fact, in Ireland one single variety of potato—named the 'Lumper'—would eventually

2   I cannot resist listing here the chapter headings of Pollan's imaginative book, which are organized around the human desire for the following: 1. Sweetness/Apple; 2. Beauty/Tulip; 3. Intoxication/Marijuana; 4. Control/Potato.

3   Before the arrival of the potato, just over one million people populated Ireland; from 1760 to 1840, the population exploded to nearly nine million.

become the seed stock of choice planted in field after field, county after county, across the island.

By the 1840s, the average adult ate between 9 and 14 pounds of potatoes per day, over the course of three meals. The 'Lumper' proved to have no resistance to *Phytophthora infestans*, the "plant destroyer," which acts like a fungus, when it arrived on the shores of the green island (or, more precisely, on the air). In the years before the Great Famine consumed the Irish people, plant disease and food shortages were common. By 1840, more than 20 diseases had already adversely affected multiple harvests. Still, no one was really prepared for the speed and destructive power of the disease known as late blight; Europe had never encountered an agricultural disaster of this scale. Because the ancestry of Europe's potatoes was so uniform— derived from two parents susceptible to late blight—fields from Belgium to Denmark, Italy to Ireland, Spain to Sweden, all were touched by the disease. One year, in the span of 4 months, the blight rode the air currents so efficiently it spread over an area of 2 million square kilometers. John Reader comments: "Neither the Vandal hordes nor the bubonic plague had penetrated Europe so deeply and so fast."

I shall pause here, for a paragraph, to bear witness to the unfortunate efficiency of phytophthora—sometimes called the world's worst agricultural disease. When we first planted potatoes on our organic farm here on the South Fork of Long Island, I was told by a conventional farmer that we were "playing with fire." My attitude, though not cavalier, was somewhat naïve. For the next 20 years, I spent countless hours in physical and mental combat with the Colorado potato beetle, but I was spared any contact with late blight. In fact, after almost 30 years of gardening and farming, I was unable to identify this disease—I had never seen it. Then in 2009, apparently introduced into the region via tomato seedling plants transported from the South to box stores in the Northeast, phytophthora did indeed take hold and spread like fire, and we were in its path. Like most other growers in the Northeast, we lost our entire tomato crop, more than 8,000 plants. And I became all too familiar with the whitish spores and blackened leaves of blight on this solanaceous plant (the Solanaceae family also includes potatoes, eggplant, peppers, and tomatillos, though I have not seen these species affected by the disease). We were spared any sight of the blight during the fine 2010 growing season, but phytophthora reappeared with a vengeance in 2011 and 2012—there is no credible explanation for it. We have always used crop rotation—as every responsible farmer must—and we have experimented with various organic spray compounds, but ultimately we are not in control of the particular strain of phytophthora that comes in on the wind. We are at the mercy of spores that can travel 30 miles (and 30 miles more . . .) from field to field. In 2009, we lost every tomato plant; in 2012, because the strain of the disease was less virulent, some plants continued to produce healthy fruit throughout the summer. Thanks to luck, along with some careful attention, timely planting, and cultivation, our potatoes have not been affected. I have learned to anticipate and to identify this outspoken funguslike disease, though

I would prefer that any future contact occurs far from any field that I choose to cultivate.

✳

IN THE PLACE of its origin, the Andes, the potato has been nourished and protected by a human culture that truly shared in the colorful evolution of this food plant. This is the meaning of plant husbandry—to cultivate a relationship with another species that assists the evolution of both. Once seed stock was exported to the fields of Europe, the human/plant relationship altered, as the thread connecting us with the natural world continued to thin. Following the devastation of late blight and various other maladies, the ingenuity of a few individuals was essential to ensure the survival of this important food source. In 1906, a former British medical doctor by the name of Redcliffe Salaman—who eventually published *The History and Social Influence of the Potato* in 1949—requested a few samples of wild potato stock from the Royal Botanic Gardens at Kew to enhance his breeding experiments. Oddly, and fortuitously, Kew's labeling system contained an error, and instead of a common wild species, *Solanum maglia*, Salaman was given a species known as *S. edinense*. When he discovered that this species contained some resistance to phytophthora, or late blight, he decided that true resistance must be found in nondomestic varieties. Salaman crossed another wild potato from Mexico, *S. demissum*, with domestic stock, and then for years he crossed and recrossed varieties with proven immunity until he had a series of highly resistant families. By 1926, 81 years after late blight spores invaded the fields of Europe, Salaman had produced sufficient planting stock to improve the viability of this South American import in the soils of another continent. Imagine the altered history of the potato, and of human history, if a diversity of species had found a way onto the ships bound for Spain or Ireland centuries earlier.

In 1938, when J. G. (Jack) Hawkes was a 23-year-old PhD candidate in botany at Cambridge, he began what would be a lifelong service to the humble potato. Jack Hawkes was chosen to join the team bound for South America during that year as a part of the illustriously named British Empire Potato Collecting Expedition. Prior to this expedition, Hawkes had visited several research stations in Russia, and while there he met Nikolai Vavilov at the Leningrad (St. Petersburg) Vavilov Institute. Vavilov himself had earlier collected samples in South America, and he urged the young student to return to England with at least 12 new species of *S. tuberosum*. After spending 8 months in the Andes traveling more than 9,000 miles, the expedition returned with a total of 1,164 specimens, and Hawkes spent decades researching and reporting on what is now known as the Commonwealth Potato Collection. *(The Potato: Evolution, Biodiversity and Genetic Resources*, published in 1990, records his experience.) Since that time, selections from this collection of wild and domesticated species have been planted out on a regular basis, providing a resource for botanists and researchers to improve seed stock of the tuber that has traversed the world (it is cultivated in 149 countries).

✣

THE CLOSE OF the spud season now arrives after Thanksgiving and extends even into December, when an enthusiastic family of farmer friends will gather to assist us in lifting the last tubers. By this time, we are 7 months on from when we first placed the seed in the furrow. We reserve a map to remind us—rows of the classic white 'Kennebec'; the Dutch yellow 'Bintze'; and the drama of the night sky in 'Purple Majesty'. I prefer to drive the tractor that pulls our potato digger—really an oversize shovel with a conveyor belt attached—that lifts the spuds, filters out earth and vegetative matter, and deposits the tubers on the surface of the soil. For hours, we move up and down the rows as potato harvesters have done for millennia (earlier models of our potato digger were once pulled by horses or oxen). Unearthed and collected in bins, a few thousand pounds of spuds are escorted to the root cellar, storage food for the community, to see us through the winter. When I stack the bins in the cellar, I am again connected with a diversity of planters and harvesters—Andean, Spanish, Russian, German, Irish, and American. I understand the seed language of potato (*batatas*), and, in gratitude, I listen for the song that filters through the dense air from Mother Earth, Mary, Pachamama.

When wheate is greene, my leaves rise
livened by branch and burr.
Now, out where ridge thaw whitens east
A thread circles for the bell point.

—from my *Book of Odes,* inspired by William Shakespeare

CHAPTER 10

Ripeness Is All

BY THE TIME OF THE summer solstice, our Amagansett fields are full of growing plants. We begin to seed directly into the soil in early April, and transplanting—of plants that were first nurtured in the heated greenhouses—will continue throughout the summer, with successions of those plants able to survive the cold April nights and the hot, humid days of July. By late June, the harvest season for the perennial crops asparagus and rhubarb is coming to a close; spinach begins to bolt; the early lettuce, radishes, and Hakurei turnips have been harvested; and the first flowers and tiny fruits begin to appear in the summer squash and zucchini patch. Zulu daisies, calendula, and typically a bed of bachelor's buttons that will reseed itself from the previous season are the first flowers to announce summer's arrival. Tomatoes and eggplant, hoping for some heat, gain stature and begin to leaf out in anticipation of August.

All of these crops require nearly constant attention, because the plant growth we prefer not to encourage—nutsedge, redroot pigweed and lambsquarters—thrives in the heat of June and July. But I also have my eye on another crop, about waist high at the solstice and turning from green to gold in the sloping field where we seeded it last October. Given our proximity to the ocean—a mile away as the crow flies—we are visited each afternoon by a sea breeze, the wind that shakes the barley (as the tune goes), though in this field it shakes the wheat. To watch the wind streak through the wheat is more than mesmerizing; it is to witness the ancient connection we have with plants, and especially with this species of the grass family, *Triticum aestivum*.

We tend to think of flowers as ornamental, or showy, though the flowers of the grass family, Gramineae (sometimes known as Poaceae), are tiny, almost impossible to see. Grasses—including both wild and cultivated species—cover one-third of the landmass of earth; only the orchid and daisy plant families comprise a larger number of species (23 species of *Triticum* have been classified). Wheat, along with rye, oats, barley, corn, rice, sorghum, and millet, are angiosperms— flowering plants. Like other plants in the grass family, wheat has narrow leaves and hollow stems, except where the leaf is attached, at the node. The base of the leaf, wrapped around the stem, is known as a sheath, above which sits an arrangement of flowers or inflorescence. On the spikelet, where the grain seed will develop, is the lemma, a scale that encloses a single flower, and inside that is another scale, the palea, that encloses the flower parts. In vast fields throughout the globe, the intricate flowers of this elegant grain move with every breeze, mostly concealed.

Grains, or cereals—after the Roman goddess of fertility, Ceres—have been integral to the advance of agriculture and civilization. The granule that matures on the stalk—a "berry," as we often call it—is full of nourishment: starches, proteins, minerals, and vitamins.[1] This is the seed of the wheat plant, and once ground or milled, its by-product, flour—for bread, chapati, pita, tortillas—sustains humans everywhere. It is the presence of gluten in the grain, made up of proteins, that allows us to knead dough and to form loaves of bread. And certain varieties of wheat, including those cultivated by early farmers, have a high protein content of 8 to 16 percent (in contrast, the other two essential grain crops that feed the world, rice and corn, contain a lower percentage of protein).

Wheat can adapt to many soils and a variety of climates, including the hard red spring wheats—also known as the "aristocrat of wheat"—that grow well in the northern regions,

---

1   The wheat berry consists of three parts: An outer layer of bran encloses the entire kernel; the germ, or embryo, is the eventual sprout of this seed; and the endosperm (83 percent of the kernel) serves as the food for the seedling.

though it cannot tolerate high heat and humidity (rice prefers the tropics, and maize (corn) grows well from the semitropics to the temperate zone). It is not surprising that a plant so adaptable and nutritious is credited with being the first crop domesticated by our ancestors about 10,000 years ago in the Fertile Crescent, at the dawn of agriculture.

The land of the Fertile Crescent spreads out from the eastern shore of the Mediterranean and includes the present-day Israel, Turkey, Jordan, Syria, and parts of Iraq. It was here that emmer wheat and another variety, einkorn, were first cultivated; ancient wild einkorn from the Karacadag mountain region of southeastern Turkey has been found to contain the same gametes (pollen and eggs) as modern domesticated wheat. By 6500 BC, emmer had found its way to Greece, and 1,500 years later, to what is now Germany. In *Guns, Germs, and Steel*, Jared Diamond, a professor of geography at UCLA, points to the balanced climate of the Fertile Crescent as the primary advantage for the growth of "big-seeded" cereals and pulses that constitute our major food sources. This climate selects for annual plants that devote their energy to producing seeds rather than woody stems. Six of the 12 major crops that feed us today derive from this region of long, dry summers and mild, wet winters.

Because the wild cereals of the Fertile Crescent were so abundant, it is likely that hunter-gatherers were harvesting, consuming, and storing the seeds of these plants long before they experimented with cultivation. Emmer and einkorn wheat, and also barley, are what are known as "selfers"—they are hermaphroditic plants that usually pollinate themselves, an advantage for early farmers learning the art of seed saving. Occasionally they will cross-pollinate, another advantage for a farmer intent on selecting the heartiest, most productive plants. The central task for the early farmer was to harvest seed from plants that did not easily shatter, for these wild cereals had evolved a protective husk, borne on a central axis known as a rachis, that would burst open at ripeness and scatter the seed for the next generation of plants. A single genetic mutation, normally occurring in a few plants, will produce a tough rachis, a trait that restricts the shattering process. The earliest farmers would gather from the most accessible plants—those that retained their seed husks. Archaeologists have estimated that after 200 years of selecting from shatterproof plants and replanting their seed, this trait would become predominant.

As the early farmers honed their skills and shifted from hunting-gathering to crop husbandry 9,000 to 10,000 years ago, much of the variation naturally occurring in wild species was lost. Within species, wild plants had evolved to mature and produce seeds at varying rates; this was a survival mechanism— plants programmed to mature over a long time span were naturally resilient, more able to resist pests and disease. Also, prior to domestication, seeds from the same or related species did not all germinate at the same time—a trait for irregular dormancy was an advantage for wild plants. Domestication changed not only our relationship with plants but the nature of the plants themselves. The long process to develop uniformity may have been a subconscious choice, but it began thousands of years ago in the wheat fields of the Fertile Crescent.

By 6000 BC, wheat had arrived in Egypt. Realizing the potential of this nutritious grain, early Egyptians became adept at building ovens and baking simple breads. As of 3000 BC, wheat had reached present-day England, Ireland, Scandinavia, Spain, Ethiopia, and India. Samples found in a granary in Assiros, Macedonia (Greece), dating from 1350 BC, have been identified by DNA analysis as the first bread wheat with a gluten content sufficient to produce yeasted bread. The dispersion method for this most adaptable of crops is clear from the archaeological evidence: Humans, ever in motion, carried the seeds forward.

✻

BECAUSE WHEAT FLOWERS contain both male and female parts (they are called perfect flowers) and therefore self-pollinate, it is difficult to produce hybrid wheat on a commercial scale—as is possible with maize. Genetically, this most nutritious of grasses is more complicated than many domesticated plants. Abdullah Jaradat, a research agronomist with the USDA, claims that wheat is "perhaps the most variable crop on earth."

In a chapter from her book *Breed Your Own Vegetable Varieties*, Carol Deppe introduces "801 Interesting Plants" to inspire a new generation of amateur breeders and seed savers. Her taxonomy includes: "scientific names, common names, families, and life styles; chromosome numbers, basic breeding systems, flowering patterns, flower types and modifications, average cross-pollination frequency, major pollen vectors, and incompatibility system information; recommended isolation distances, seed yields, location in the USDA-ARS National Plant Germplasm System, and references." (A biologist's list!) For *Triticum* she lists 15 cultivars, ranging from Persian wheat to Polish wheat, including emmer, einkorn, and spelt. Though most plants are classified as haploid—they have one copy of each kind of chromosome—einkorn wheat is a diploid plant (possessing two copies of each kind of chromosome).[2] *Triticum aestivum*, or bread wheat, is a hexaploid species—with six copies of each gene—and worldwide, it is the most cultivated; spelt, grown in lesser quantities, is also hexaploid. To put it another way, a hexaploid possesses 16 billion pairs of DNA—40 times more than rice, 6 times more than maize, and 5 times more than humans! Durum wheat, a tetraploid species (with four copies of each gene, like the ancient emmer wheat) used for pasta, bulgur, and semolina, is the second most cultivated in the world. Unless one is a breeder, this level of detail is likely to be confusing, but it does reveal the complexity of one of the world's staple foods.

In the centers of diversity and in fields cultivated by indigenous people, landraces have long been used for breeding and replanting to maintain diversity within a crop such as wheat, although this practice has been absent from conventional

---

2   The term *ploidy* refers to the number of haploid genomes in a cell, plant, variety, or species. Polyploid plants have more than two complete sets of chromosomes—triploid, tetraploid, hexaploid, etc.

agricultural systems. A landrace is a population of plants that has evolved over a long period of time, subject to natural and human selection, that is heterogeneous (displaying significant variation) but adapted to the local system. Landraces *must* be protected and kept in circulation—there is no other way to say it. (Eli Rogosa, of the Heritage Grain Conservancy, who is breeding rare landraces today—emmer, einkorn, Ethiopian purple, among others—is insistent that these cultivars have the best chance to cope with changing climatic conditions, and nutritionally they are superior to modern varieties.)

In ancient wheat cultivars—unlike the modern varieties— the grain kernel stays encased in the glume, or hull, so that a more rigorous threshing is required to free the edible portion of the grain from the hull. Despite this, emmer has been continuously cultivated for centuries—in the Midwest, it has been widely used for livestock feed, and in Italy, where it is known as farro, it is traditionally used for pasta. German immigrants from the black soil belt of Eastern Europe, who settled in the high plains of North Dakota and Manitoba in the late 19th century, brought with them numerous strains of emmer, and descendants of these farmers have saved the seed for generations. Steve Zwinger, a farmer and researcher at North Dakota State University, has been trialing old landrace varieties for more than 10 years and is largely responsible for the present resurgence of this ancient wheat. Jack Lazor, a grain farmer in northern Vermont, in his classic book *The Organic Grain Grower: Small-Scale, Holistic Grain Production for the Home and Market Producer,* observes: "Emmer's greatest asset is that no one has tampered with its genetics for the past ten thousand years."

Formal wheat breeding began in the 19th century and intensified in the mid-20th century with the rise in availability of artificial fertilizers. The increase in nitrogen applications—to promote greater yields—led to heavier seed heads of wheat that tended to fall over in the field (commonly called "lodging"), so breeders were forced to search for shorter varieties. In the 19th century, the Japanese had developed several "dwarf" varieties, and these impressed US commissioner of agriculture Horace Capron when he visited Japan in 1873. These dwarf varieties were crossed with strains from other countries. Decades later, American agronomist Norman Borlaug incorporated the variety known as 'Norin 10' in his field trials in Mexico. By 1963, having found success with increased disease resistance and soaring yields, 95 percent of Mexico's wheat crop was planted out with Borlaug's new varieties.[3]

In 2010, world production of wheat hit 651 million tons, ranking it third among cereals; maize production yielded 844 million tons and rice 672 million tons. More acreage is

---

3   Some experts, including William Davis, MD, author of *Wheat Belly*, claim Borlaug's high-yield, semidwarf wheat and varieties derived from it created a very different food substance that has had measurable negative impact on human health. Davis's research suggests that the presence of a protein in modern wheats, gliadin—a subcomponent of gluten—acts as an appetite stimulant, and that when the modern semidwarf wheat reached the supermarket shelves in the United States around 1984, there was an immediate increase in calorie intake of 400 to 800 calories per person per day. He and others advocate removing modern wheat from the diet to reduce chronic weight problems and related diseases such as diabetes and heart disease.

planted in wheat than any other crop—nearly 600 million acres worldwide—and world trade in wheat in dollars tops all other crops combined. China currently produces about one-sixth of the world's wheat, followed by India, Russia, the United States, and France. The United States, the European Union, Canada, Russia, Australia, and the Ukraine are the primary exporters of wheat, while the primary importers are Egypt, several of the EU countries, Brazil, Indonesia, Algeria, and Japan. Despite the importance placed on yields in the last 70 years, productivity remains highly variable.

Cary Fowler and Pat Roy Mooney label the problem of genetic erosion first noticed by American botanist Jack Harlan in the 1930s as "an unending avalanche" following the establishment of crop-breeding institutes throughout the world (the research station in Mexico was one of these). By 1970, nearly all of the indigenous wheat varieties in Greece were gone, and by 1976, 44 percent of all land planted in wheat, worldwide, was planted using the new, miracle varieties. One story recounted in their book *Shattering* emphasizes the importance of preserving indigenous varieties and of protecting genetic viability. In 1948, Harlan, on an expedition to Turkey, collected an uninspiring wheat cultivar that he designated with the number 178383 in place of a name. "It is a miserable looking wheat, tall, thin-stemmed, lodges badly, is susceptible to leaf rust, lacks winter hardiness . . . and has poor baking qualities," he wrote about this homely specimen. Yet when a stripe rust epidemic hit the northwestern states, this miserable-looking wheat, resistant to four races of stripe rust, saved the day. For many years, 178383 was used in breeding programs across the northwestern region of the United States.

As of late 2013, no genetically engineered wheat had yet been approved for planting anywhere in the world. However, in May of that year, an Oregon farmer discovered evidence of GMO contamination in his conventional wheat field. Monsanto had tested GE varieties in 17 states between 1998 and 2004, and the variety found in the Oregon field matched one of those tested by the biotech company. When South Korea and Japan suspended US wheat imports, Ernest Barnes, a Kansas wheat farmer, filed a lawsuit against Monsanto, citing their "gross negligence," and claiming his "livelihood is now at serious risk." Monsanto had no explanation for the contamination. Calling the incident "serious," the US Animal and Plant Health Inspection Service (APHIS) immediately initiated a formal investigation.

Meanwhile, industrial systems developed for greater efficiency, and to increase production of commodity wheat (and other cereals), have severely compromised some of the world's great soils. Annually, the use of the moldboard plow—John Deere's 1837 invention that cuts deep into the soil and inverts it—over vast acreage alters the aggregate structure of the soil, making it vulnerable to wind and water erosion. In an article titled "Tilth and Technology," Peter Warshall writes: "Since 1950, about one-third of American cropped land has had to be abandoned because of erosion problems." Numerous studies have concluded that when the true cost of inputs is considered, a monoculture system is in reality less efficient than smaller, diverse models.

❦

IN *STOLEN HARVEST*, Vandana Shiva writes about the theft of *kanak*, or gold, as wheat is called in northern India. The wheat economy is vast in India—a nation of 1½ billion people—employing millions "while ensuring the availability of fresh, wholesome, sustainably produced and processed, inexpensive food." The system is decentralized and based on local producers, processors (*chakki wallas*, or local flour mills), and traders (*artis*). Consumers buy wheat at a local shop (*kirana*) and carry the grain to the nearest chakki walla (over 2 million exist in India) to be ground into flour. The Indian people cherish freshness and flavor; Shiva comments that "less than 1 percent of the flour consumed in India comes from packaged brands."

This complex system, developed over centuries, is judged by agribusiness analysis to be underdeveloped. One industry report claims that modernizing the wheat economy could add 50 million jobs. Vandana Shiva reports a different story, calculating the 100 million livelihoods that may be lost as the result of a takeover motivated by profit, including 20 million to 30 million farmers, 5 million chakki wallas, 5 million artis, 3.5 million kiranas, and millions of households engaged in local production.

❦

AFTER YEARS OF increasing our own farm's vegetable and flower varieties, so that they now total more than 500, we have started to include wheat in our plantings, inspired by two of my former apprentices. We have been growing grains—as cover crops—for more than 20 years, but to grow grains for food is another adventure. In the 21st century, this is an uncommon crop on the eastern end of Long Island, though a landscape dotted with windmills suggests that the story has changed over the course of a few centuries. In the 17th century, Long Island actually served as the breadbasket for the region, where seed brought over by the Dutch and the English (wheat is not indigenous to North America) was planted productively—for a while. After the sandy soils were exhausted by repetitive cropping, wheat production moved to the Hudson Valley.

When Amanda Merrow and Katie Baldwin founded Amber Waves Farm in Amagansett, they developed a passion for this ancient crop, and their intuition, and perhaps my mentorship, guided them (it is always wise for each small farm to find a niche). For the first couple of years, we helped them to prepare the ground—we plowed and dragged the disc harrow to loosen the soil until they could purchase their own tractor, and they experimented by planting a hard red winter wheat, 'Expedition' (planted in the autumn), supplied by the most knowledgeable upstate New York cereal growers we know, Lakeview Organic Grain, and 'Frederick', a soft white winter wheat. In 2011, the year in which the grain harvest in the Midwest was devastated by drought, we each planted the red winter wheat available to us—a hard red variety known as 'Arapaho'. Heather Darby, a University of Vermont Extension agronomist and one of the persons responsible for the revival of grain growing in the Northeast, has had success with this variety in trials conducted in the Champlain Valley. ('AC Morley', 'Maxine', 'Harvard', and 'Warthog' are all impressive varieties in terms of yield and grain quality.) Depending on local climatic conditions, the wheat grower has the choice of selecting a winter or a spring wheat variety; winter varieties mature over a 9-month period, spring varieties over a 4-month period. (Color is determined by phenolic compounds contained in the bran layer of the grain; hard wheats, used for bread making and brewing, are higher in protein, and soft wheats are preferable for pastry flour.)

The stages that follow planting require some research, endless practice, ingenuity, and cash. There is not a lot of time for reflection; at a precise moment, the crop must be cut and harvested—"ripeness is all"—then cleaned and stored. This requires labor, tools and/or machinery, and a mind-set geared for efficiency. Wheat berries, the edible seeds of the wheat plant, are popular today as a health food used in cereal or salad; to grind flour, now that the windmills are retired, the grower must double as the miller.

So the young farmers at Amber Waves invested in a vintage combine (to harvest the wheat, one must separate the grain from the chaff and then winnow the grain from the seed head—this can be accomplished by human hands wielding a scythe or by a machine). Their machine is a 1956 John Deere, lovingly held together by wire and duct tape and urged to perform by Pete Ludlow, the practical and musical scion (he's a superb organist) of a family that has farmed on the South Fork of Long Island for 250 years. What we harvest we eat: At

Amber Waves, they include a loaf of bread in each CSA pickup, and at our farm, we now distribute wheat berries to our CSA members at every harvest day after the combine has rumbled through the field of grain.

Stephen Jones, a wheat breeder and the director of the Washington State University Research and Extension Center in eastern Washington, has spent the last 30 years focused on the development of a supply chain that involves growers, millers, and bakers. He says: "Grains are vital, vital for our food sovereignty and our food security." Dedicated to reviving local growing networks and to encouraging sustainable farming practices, he has been tireless in the effort to decentralize what is a vast and dominant industry, global in its reach. As part of his breeding program, Jones has been working with others to develop a perennial wheat; Wes Jackson, one of the leaders of the ecological farming movement, founded the Land Institute in Kansas specifically to focus on the breeding and production of perennial grains. Jackson estimates that annual cropping systems—*the* standard agricultural paradigm—release 40 percent of the total carbon held in the soil (precisely at the time when we are working to sequester carbon in our soils), and he maintains that perennial systems "are more resilient, ecologically restorative, and economically profitable than annual monocropping systems that hardly mimic nature at all" (in the words of Fred Kirschenmann).

For many years, Vermont served as the breadbasket for all of New England. As the agricultural commodity system developed and grain production shifted to the West, the Northeast breadbasket emptied—by 1930, Vermont and Maine produced no wheat at all. Now, with the support of MOFGA (Maine Organic Farmers and Gardeners Association) and the NOFAs (Northeast Organic Farming Association—each New England state has a chapter), grain production is again on the rise in the region. The intention of multiple projects, such as NOW (Northeast Organic Wheat) and Farm to Bakery, is to strengthen the regional food system through the production of grains for human consumption.

"Grains are vital . . . ," Stephen Jones observes, and we should listen to a creative breeder with more than 30 years' experience in the industry and university extension. In reference to the growing movement that fosters collaborative relationships among growers, millers, and bakers, he offers some promise, "It is hopping now!"

Our quixotic, natural, great sky
rains down, yields its boundless form
to the furrows in the weave of your time.

And you name the darling mischief and the hope
of each variety as it holds you, where it germinates,
to the flood plain of your own kernel spirit.

So will braids of garlic whiten, whiten.

from "Quail Hill Farmers," Megan Chaskey

CHAPTER 11

Trust in *Allium sativum*

GARLIC CAST A SPELL on our inquisitive species more than 5,000 years ago, somewhere amidst the soil and rock of the mountains of south-central Asia. Cultivation is thought to have originated in the foothills of the Tien Shan, Pamir, and Kopet Dag ranges; what is now known as the "garlic crescent" extends along the present Russian border with China, Afghanistan, and Iran. In fact, some scholars propose that as far back as 10,000 years ago, hunter-gatherers were already cultivating a wild species known as *Allium longicuspis*. A word for garlic turned up in the oldest known written language, Sanskrit, and friendship with this allium can be traced back through all of the early civilized cultures. Contemporary lovers and planters of garlic have an ample store of historical lore to defend what some might label as fanaticism. I know of no other crop that attracts such a (hmm) colorful cast of supporters.

As a cultivar, this allium has many admirable traits, including the ability to thrive at almost any elevation and under a diverse range of conditions. Garlic can be admired for its taste, a willingness to enhance almost any cuisine, and for its medicinal properties. I will not tolerate casual criticism of this versatile plant, though I have tried, consciously, to be more discreet in conversation concerning *A. sativum* var. *ophioscorodon*. As a food, garlic is nearly ubiquitous, but as a plant, it is little understood by those who consume it.

I first planted garlic in the cliff meadows above Mount's Bay, Cornwall, the fertile ground where I was initiated into gardening. Until I encountered a few handsome bulbs in a friend's Cornish garden, I assumed that this food plant flourished only in Italy, or France, or California. Although it was challenging to grow food crops in the Cornish hillside meadows—rabbits, voles, and badgers often preceded me to the harvest—the garlic I planted there matured to a fine size and flavor. My luck with this most adaptable of alliums continued for almost 30 years until I, or rather, the cloves I tended, had a run-in with representatives of the species garlic bloat nematode. Though the nematodes themselves were undetectable by my watchful eye, the damage certainly was apparent—we lost over one-half of our crop in the summer of 2010.

Garlic is an asexual plant; therefore, it does not produce true seed (although, just to confuse us, some breeders have managed to encourage seed production). The plant normally reproduces by means of a vegetative process, the same process performed by daffodils and narcissi. To plant a crop—properly done in late fall, 6 weeks before hard frosts—one is required to break apart the garlic bulb and to separate the cloves; each clove becomes, in effect, a seed. For an unknown reason, the hardneck varieties of garlic produce a shoot that rises from the center of the plant, above the leaves, 8 months after being planted. This is commonly called a scape, though sometimes, at farmers' markets, the term *green garlic* is (incorrectly) used. If left to mature, a neat little package resembling the seed head of other plants in the allium family will form at the end of the twirling scape.[1] Like most growers, we snap off the scape—in itself, full of garlic flavor—to encourage increased growth in the bulb.

The ritual of planting is quite simple: Press a clove into well-prepared soil, say to a depth twice that of the clove, proceed

---

1   Carol Deppe, an Oregon plant breeder and a biologist, defines the twirling scape as a "flower stalk."

to cover the crop with mulch (shredded leaves do nicely), and then practice patience. Although garlic will add growth on warm days in winter and will proudly proclaim "Spring!" while most plants are still in hibernation, the mature bulbs will not be ready to harvest until July. For a so-called annual plant, garlic is rare; the gardener will visit this allium often, through the winter, spring, and early summer, before he or she lifts the mature bulb from the ground, 8 months after planting.

In the early days of our CSA farm, we discovered the superior taste of hardneck or topset garlic, *A. sativum* var. *ophioscorodon*, and after many conversations we discovered that other growers in the Northeast had found success with these varieties. I discovered some impressive bulbs in a local health food store, and this led me to the proprietor of Rattlesnake Ranch, located in the high country of Idaho. He was a "caraktur," as my Cornish gardening mentor would say, a personality trait common among the serious students of *sativum*. For many years, we purchased seed (rather, bulbs chosen for planting) from Rattlesnake Ranch, and each year we saved a little more of our own. Garlic has a distinct tendency to naturalize, or to adapt to a particular soil, especially if grown out year after year at the same farm or garden (while practicing rotation from field to field, or plot to plot). Planted in our fine silt loam, garlic bulbs tend to take on a peachy coloration; the varietal name we have chosen is 'Amagansett Peach'.

As happens when a farmer finds success (and superb flavor), year by year we increased our planting stock and our commitment to saving and sourcing good seed. Twenty years ago, we planted

25 pounds of seed; then we increased the amount to 75 pounds, then 200, 400, 650, 800 pounds, or somewhere in the neighborhood of 45,000 cloves. Remember, each clove has the potential to mature into a full bulb. Most bulbs of our favorite hardneck varieties contain five or six cloves—so when we separate the seed garlic and plant one clove in the fall, we expect to harvest a bulb in July that has multiplied by five or six times, a very decent rate of return for a farmer. Each year, as our community farm has matured, so has the garlic harvest, and we have increased our autumn plantings to provide extra for seed, planning at least a year ahead. Now, after distributing garlic (for food) throughout the summer and into the autumn, we hold back enough—700 to 800 pounds—to plant for next year's crop. Once we discovered how flamboyant and useful the scapes could be—a brilliant invention of the hardneck varieties—we were really hooked.[2] No other crop has the unique flair of *ophioscorodon* nor the resistance to flea beetles, cutworms, borers, various blights, even to erosion. Try to wash it away once those tenacious root fibers have gripped the earth!

Our good luck extended—not without a lot of work—for a full 20 years. Then, in May 2010, on a walk through the rows, I began to notice some significant yellowing in the leaves about 2 months prior to the expected harvest. I had encountered something like this before (can a farmer or gardener ever expect 100 percent success?), but in the past the dieback had

---

2   What does one do with 45,000 scapes? For years, they enriched our compost heap. Wiser now, we commission a local commercial kitchen to produce cases of garlic scape pesto and multiple jars packed with pickled scapes.

been minimal, isolated. This time, when I pulled the problem stalks to check for bulb formation, the seed clove and potential bulb had turned to paper. There was nothing for the expected harvest except a collapsed cone of wrappers. "Impossible," I thought. "Garlic is our charmed cultivar; a fine harvest is assured."

I called a few allium enthusiasts, consulted my library, and increased my visits to the garlic patch. Basal rot (catchy name!) seemed to be a possibility, but I was not convinced. What had been isolated continued to migrate through the field—multiple stalks yellowed and sagged. Not until I finally called the founder of the Garlic Seed Foundation did I discover the probable cause of this unforeseen dieback: the garlic bloat nematode.

Nematodes, or roundworms, are the most numerous multicellular animal on earth, which has led biologist E. O. Wilson to proclaim: "It's a nematode world." Nematodes have been characterized as a tube within a tube—they are the most diverse phylum of pseudocoelomates (20,000 species are included in the phylum Nemata), and a handful of soil contains many thousands of this microscopic worm. Not all nematodes are agricultural pests; some are, in fact, used in biocontrol measures against certain insects. Those who study *Ditylenchus dipsaci* have access to decades of research and the support of the International Federation of Nematology Societies.

Here on Long Island, the golden nematode has plagued potato fields for decades. Once discovered to be infected, a field is certain to be condemned, closed until further notice. With care, some good fortune, and because of calculated rotations, our fields—all former potato ground—have been free of this particular roundworm throughout our tenure here. Until the spring of 2010, I had never heard mention of the garlic bloat nematode, nor had any of my grower friends.

I have always been taught that it is good practice to reinvigorate fields with fresh planting stock, assuming the source is a reliable one. At the time, we had increased our own seed garlic supply to about 300 pounds, and yearly we would trial several new varieties, searching for diverse flavor. At the appearance of the nematode, the damage at first appeared to be random—some of all varieties were fading in the field.

By the first week in July, when the crop was ready for harvest, we determined that more than half was lost. What was left, thousands of bulbs, was still good for eating, but we could not save seed for future planting; the risk is too great to reuse bulbs harvested from an infected field. Now we were faced with the next dilemma—how to locate seed for the fall planting, next year's crop?

Although many growers I called had been spared, and many were unaware, seed indeed was scarce. By reading and listening, we learned that several large-scale suppliers had disseminated infected seed throughout the Northeast. Several seed catalog companies, unaware of the problem, had also passed on the seed weakened by the nematodes. Digging deeper, we found that outbreaks of the bloat nematode were not uncommon in Canada in recent years, and, yes, we had imported some (fine looking) 'German White' from a seed supplier I had trusted. Now, isolated in an allium patch, we had unwittingly inherited a dilemma. The unsettling experience of losing much of my favorite crop ("just when I thought I was safest . . . ," as

Whitman intoned) raised questions that, as a grower of plants, I had never before asked. When I attended a seminar hosted by our land grant extension service, I was surprised to learn that the US Department of Homeland Security—called in because of the commercial transfer of material across the US/Canadian border—was also asking questions regarding the traffic of garlic seed.

With the help of our grower friends, we managed to find enough excellent hardneck seed to plant our next crop at the end of October—reduced by 20 percent from the previous year. We also ordered a variety of mustard seed for use as a cover crop in the former garlic field. We have joined as partners in this bioremedial research; the roots of certain brassicas, especially mustards, release glucosalinates that produce a chemical reaction in the soil unpalatable to nematodes. I wish the round-worms no harm, but I would prefer that they find another patch of earth to inhabit.

※

THE FOLLOWING YEAR of tending and harvesting restored our courage. Wilt did not visit in May; instead, we were delighted in June by the improvisational dance of thousands of garlic scapes. After distributing bulbs throughout the summer, to the satisfaction of our farm members, we saved a few hundred pounds for replanting. Given the rising cost of seed stock and the threat of pests impossible to see, we decided to follow the example of various other growers and lovers of garlic. Now, each year, we increase our autumn planting to produce all of our own seed for the following year. After decades of cultivating a tender relationship with this charming allium, I remain unwilling to allow the bloat nematode to disturb our careful serenity. We had a lucky run for more than 20 years, good fortune we are working to revive. I am still under the spell cast when the first wild garlic willed human hands to separate cloves to ensure a sulfurous progeny. I am a devoted seedsman, and I trust in *sativum*—the acrobatic art of the plant, the delicate and zingy flavor, and, foremost, this most defining character trait: adaptability. Listening, I hope to learn from it.

Ay, alma mia,   *Oh yes,*
hermoso   *the planet*
es el planeta. . . .   *is sublime!*

—Pablo Neruda, translated by Ken Krabbenhoft

CHAPTER 12

Seed Freedom

ONE OF MY FAVORITE NOVELS, *Joy of Man's Desiring*, a kind of hymn to nature written by the French author Jean Giono—and translated by Katherine Allen Clark—begins like this (I have to quote the first pages in their entirety to evoke what the author Henry Miller calls "the mythical domain"):

> It was an extraordinary night.
> The wind had been blowing: it had ceased, and the stars had sprouted like weeds. They were in tufts with roots of gold, full-blown, sunk into the darkness and raising shining masses of night.

Jourdan could not sleep. He turned and tossed.

"The night is wonderfully bright," he said to himself.

He had never seen the like before.

The sky was vibrating like a sheet of metal. You could not tell what made it do so because all was still, even the tiniest willow twig. It was not the wind. It was simply that the sky came down and touched the earth, raked the plains, struck the mountains, and made the corridors of the forests ring. Then it rose once more to the far heights.

Jourdan tried to awaken his wife.

"Are you asleep?"

"Yes."

"But you answer."

"No."

"Have you seen the night?"

"No."

"It is marvellously bright."

She did not answer but heaved a deep sigh, smacked her lips, and shrugged her shoulders as if to rid herself of a load.

"Do you know what I am thinking?"

"No."

"I feel like going out to plough among the almond trees."

"Yes."

"The piece out there beyond the gate."

"Yes."

"Over toward Fra-Josépine."

"Oh yes!" she said.

Again she made that movement of her shoulders and finally lay flat on her stomach, her face buried in the pillow.

"But I mean now," said Jourdan.

He got up. The floor was cold, his corduroy trousers glacial. There were flashes of this night everywhere in the room. Outside could be seen, as plainly almost as in the daytime, the whole plateau and Gremone Forest. The stars were everywhere.

Jourdan went downstairs to the stable. The horse was asleep on his feet.

"Ah," he said, "you know anyway. You didn't even dare to lie down."

He opened the wide stable door. It gave directly onto the fields. When the light of the night was seen like that, without a pane of glass between, one suddenly realized its purity, one perceived that the light of the lantern, with its kerosene, was dirty, and that its blood was choked with soot.

No moon, oh, no moon! But it was as though one were beneath glowing embers, in spite of the onset of winter and the cold. The sky smelt of ashes. It had the odour of almond bark and of the dry forest.

Jourdan thought that this was the moment to use the new metal plough. Its limbs still wore the bright blue of the last fair; it smelt of the store but it had a willing look. This was the time to use it if ever there was one. The horse was awake. She had come to the door to look out.

There are on the earth moments of great beauty and peace.

I begin with this passage because, more than any other, it evokes the very bodily joy of turning the earth, and of the calling—from the air, the stars, the soil—to do so. I am reminded of a familiar feeling: of resting on the handle of a spade and looking back on the deep brown earth, newly spaded over, in a garden on Boar's Hill, Oxford. And of lifting the rich soil of Penwith, Cornwall, with each dip of the long-handled shovel, dropping the handle onto my thigh just above the knee to throw the soil up the steep incline, working as a plow across and back the contour of the compact cliff meadow. And of bending down to lift a palmful of Amagansett silt loam from the furrow after the chisel plow has first opened the soil in April, noting the shine on the steel shanks.

To raise a crop with some success, and to husband a plant to produce ripe fruit, a gardener or farmer needs to know the soil (in changing weather), the character and adaptability of the seed, and how to prepare a seedbed. The soil is mineral matter, and when nourished, it is alive with micro- and macro-fauna and seeds native to the place—deposited by previous plant species, by wind, by passing birds. A farmer comes to sense the pulse of life within the soil, and in spring the song rises up, calling for seeds. It is time to invest in the earth, to lower the plow, to introduce the soil again to the air, to loosen and ply apart the contraction of winter. The seedbed should be clean and fine to ensure germination, organic matter given over, folded in from the previous season to enrich the soil that receives the seed.

"The soil is the seat of abundance on earth," wrote Peter Farb, journalist and author of *Living Earth*, "a massive machinery for keeping the chemical stuff of the planet in constant circulation. There can be no life without soil, and no soil without life—they are inseparable." For three-quarters of geologic time, soil did not exist—our planet was made of rock and water—and it is highly unlikely that the angiosperm revolution could have been born on stone. But it likely was born on soil—a mantle of minerals laid down after millions of years of movement, all the elements combining to create the original (archetypal) seedbed. The geologist Nathaniel Southgate Shaler wrote: "All soil is rock material on its way to the deep."

There is an unwritten pact that every farmer acknowledges, even if it is not always possible to attain—to leave the soil in a better condition than when he or she inherited it. This makes practical and economic sense, of course, but more important, the one who sows and cultivates is linked, body and mind, with the natural cycles of regeneration and fertility. To build good health in a soil is not a simple task, but we have thousands of years of experience to draw on, and, so far, mentors in each succeeding generation.

Recently, while preparing a talk on resilience in agriculture,

I was led back again and again to one word: *humility*. I have read that this word derives from the same root as the word *humus*, meaning earth or ground. Humus is the stuff of the soil that is left after fungi, microbial life, insects, and worms have broken down layers of leaves, branches, stems, straw, and stone—"the dark brown or black substance resulting from the slow decomposition and oxidation of organic matter on or near the surface of the earth . . . " my *Oxford English Dictionary* reports. The decomposition process—which provides the seed-bed for our food—releases minerals and carbon dioxide (energy!) to be absorbed by plants.

Sir Albert Howard, the mentor for all cultivators on the topic of soil health, taught that by returning animal wastes to the land, we can ensure good humus formation. Humus, as a substance that builds health, Howard says, is "nothing short of profound." Ultimately, we know very little about this soil substance, but the key factor is that humus acts as a reserve of nutrients until—assuming all other variables line up—plants are ready to absorb the gift. To build soil fertility, one must maintain appropriate levels of humus, and Howard proclaimed this would be "the basis of the public health system of the future." Britain's Prince Charles, who for more than 40 years has been promoting a more sustainable approach to agriculture, advised in his speech "On the Future of Food" that we need "a form of agriculture that does not exceed the carrying capacity of its local ecosystem and which recognizes that the soil is the planet's most vital renewable resource."

Perhaps you have heard that each handful of soil is alive with more than a billion microorganisms. I consider myself fortunate—because of my daily work—that I have abundant opportunity to encounter the ordinary and the miraculous as it occurs in the mantle of earth under our feet. I inherited a soil at our farm that was depleted by years of monocrop farming and the use of synthetic fertilizers. Each year, following an application of green manures (leguminous cover crops) and compost, the soil structure improves to the touch, and the plants that we introduce to this soil respond with profuse leafage and robust fruit.

Our science textbooks assure us that we have a reasonable understanding of the complex interactions that occur within any soil, and we are adept at naming the macrobial creatures that occupy the earth: beetles, springtails, spiders, millipedes, centipedes, wood lice, slugs, snails, and earthworms. Though we now recognize the value of microbial life in soils, and of the symbiosis that occurs between mycorrhizal fungi and the roots of plants, this still remains the undiscovered country. As seeds germinate and plants mature and decay, all other life-forms in the soil also flourish and recede. Life itself is the process; soil is much more than a by-product, it is the fecundity of the planet: *Every force evolves a form.*

The great soil scientist Hans Jenny always remained humble before his subject. He recognized soil as a work of art: "I have seen so many delicate shapes, forms, and colors in soil profiles that, to me, soils are beautiful." Fred Kirschenmann, the North Dakota grain farmer–philosopher, speaks of the need to "invoke a cultural memory of caring" for the soil in a particular place.

Humus formation, the making of soil, is an extraordinary

process—in an acre of ground, earthworms alone may pass
40 tons of earth through their bodies—and worm castings are
basically packaged fertility. Castings, neutralized by carbonate
of lime secreted by an earthworm's calciferous glands, contain
5 times the available nitrogen, 7 times the available phosphorus,
and 11 times the available potash of any other substance in the
top 6 inches of soil. Aided by the digestive process of earthworms, nutrients in the soil can be more easily taken up by
plants. As they move through the soil, worms do the work of
our plows, loosening the subsoil and creating an aggregate that
plant roots (and farmers) prefer. The French ecologist André
Voisin links the civilizations of man with the distribution of
earthworms; at the birth of agriculture, the valleys of the Indus,
the Euphrates, and the Nile were blessed with soils organically
improved by lumbricid earthworms. Marcel B. Bouché, in a
foreword to Dr. Kenneth E. Lee's book *Earthworms*, writes:

> If we compare, for example, the significance accorded
> to ornithology and the multitude of birdwatchers
> studying about one kilogram of birds per hectare,
> with the extremely limited number of research work-
> ers' interest in the hundreds of kilograms or tons per
> hectare of earthworms, we must conclude that our
> knowledge of ecosystems is fundamentally distorted
> by our above-ground, visual perception of nature and
> our ignorance of life below-ground.

It is certainly time for stewards of the soil to acknowledge
these other "legislators of the world" and to cultivate a sensitivity

and an awareness of the web of relationships at the heart of farming and preserving soils. In Wendell Berry's words:

> The proper business of a human economy is to make one whole thing of ourselves and this world. To make ourselves into a practical wholeness with the land under our feet maybe not altogether possible—how would we know?—but, as a goal, it at least carries us beyond hubris. . . .

꽃

VANDANA SHIVA, THE INDIAN environmentalist and activist who received the Right Livelihood Award (the alternative Nobel Peace Prize) in 1993, speaks of a mountain farming system practiced in the Garhwal Himalaya known as *baranaja*, or "12 seeds." The seeds of 12 or more crops are mixed together and distributed at random over a field that has been "dressed" with manure. We have experimented with this style of sowing on our own farm, and the harvest is always fruitful; we learn from it, about what a plant requires and prefers. The demands of the market limit our exploration, but I suspect that an ecosystem benefits from the openness and freedom of this style of sowing and cultivation. In the baranaja system, crops are chosen after many years of sowing and seeing: What are the symbiotic relationships between plants, and between plants and the chosen soil? In this system, plants are thinned and replanted to preserve a working diversity so that the harvest is beneficial both to the sower and the soil. The knowledge gained from observation can lead to greater yields and an understanding of seed and soil that is the basis for what we have labeled "sustainability." The practice of baranaja not only preserves diversity but also what Shiva refers to as food democracy. She writes that Indian women use up to 150 various species of plants for food, feed, and medicine, including countless uncultivated plants and what industrial agriculturists would term "weeds." "What is a weed for Monsanto," Shiva says, "is a medicinal plant or food for rural people."

For the people of India, one crop—rice—is identified with *prana*, or life breath. Rice has long been celebrated in folklore, song, and poetry, and in this passage from the Amritabindu Upanishad, this basic food grain is associated with knowledge of the self:

> Knowledge is twofold, lower and higher.
> Realize the Self; for all else is lower.
> Realization is rice; all else is chaff.

Over the centuries, Indian farmers developed more than 200,000 varieties of rice, bred to survive in changing climatic conditions in a diverse range of soils. Shiva points out the absurdity of a Texas-based company being granted a patent (in 1997) on one of the most refined of rice varieties, basmati. In fact, the basmati patented by RiceTec could not exist without the labor and innovation of generations of Indian farmers—this "new" rice was bred from Indian basmati crossed with semidwarf varieties. Neither the breeding method used by RiceTec nor the resulting rice variety is in fact novel, though the patent was issued on that basis. Because the livelihood of hundreds of thousands of farmers is now at risk [Shiva calls this kind of action "biopiracy."]: "The perverse system that treats plants and seeds as corporate inventions is transforming farmers' highest duties—to save seed and exchange seed with neighbors—into crimes."

In 1998, on the 68th anniversary of Gandhi's mobilization of the Indian people in the salt satyagraha ("the struggle for truth"), more than 2,000 groups joined together to proclaim a bija satyagraha, "a noncooperation movement against patents on seeds and plants." This movement is an example of peaceful noncooperation, an integral part of the democratic tradition of India; in this case, its actors were refusing to be dictated to by multinational corporations or to accept the rules and judgment of the World Trade Organization. As Shiva explains: "The bija satyagraha is an expression of the quest for freedom for all people and all species, and an assertion of our food rights." Indian farmers and environmentalists also have organized a cooperative movement for the purpose of saving seed—Navdanya. Those who participate use ecological farming practices—"ahimsic krishi," or nonviolent agriculture, based on compassion for life—and pledge to save and share the seeds that are viewed not as property to be patented but as gifts from their ancestors.

In 2006, at the Terre Madre gathering in Turin, Italy,

Vandana Shiva and others met in order to prepare a Manifesto on the Future of Seed.[1] She was joined in this task by Jerry Mander, Miguel Altieri, Wendell Berry, Andrew Kimbrell, Carlo Petrini, Alice Waters, and other supporters from countries scattered throughout the globe; following the release of the Manifesto on the Future of Food in 2003, there was a recognition by the International Commission on the Future of Food and Agriculture that it was imperative to address the increasing threat worldwide to seeds and biodiversity. Carlo Petrini, the founder of Slow Food, remarked that "the strong seed of Terre Madre is the practice of the local economy," and that "the vision of local economy is not an archaic one, but an extremely modern vision." The written document is intended to inspire a new generation of food activists to engage in the business of seed conservation and rescue.

The Manifesto on the Future of Seed states that the patenting of seeds and new technologies is transforming what has always been held as a commons—a free, local exchange among farmers. Seed saving and preserving seed for the following season has been the accepted practice for thousands of years, and often this has been the domain of women. As the local knowledge of preserving seeds erodes, the economy that supports small farms erodes, and with it the security offered by a traditional food culture—and the importance of women to that culture. The nutritional impact

---

1   Terre Madre is a worldwide gathering of food communities hosted by the Slow Food network. In October 2004, more than 5,000 farmers, breeders, beekeepers, and chefs from 130 countries came together to meet in Turin, Italy. In 2006, more than 1,200 food communities were represented at Terre Madre.

of industrial systems on rural communities and the poor has been devastating.

A system that encourages the predominance of intellectual property rights (IPRs), in which seeds protected by patent must be repurchased by the farmer each year, restricts the rights of millions of smallholders (individual farmers). Independent, sovereign nations are called on by the Trade-Related Intellectual Property Rights (TRIPS) agreement to create IPR systems that support the dominance of global corporations.[2] The developed countries of the North have continued to undermine the International Treaty on Plant Genetic Resources for Food and Agriculture, an agreement intended to mitigate the excesses of the exchange of seed under the IPR system. As public funding for research and conservation of seeds has diminished, private corporations have gained more and more control of breeding and distribution; diversity, ignored, suffers. Under the heading "Privatization of Seed," the document reads:

> The transformation of a common resource into a commodity, of a self-regenerative resource into mere "input" under the control of the corporate sector, changes the nature of the seed and of agriculture itself. It robs peasants of their means of livelihood,

2   Following the founding of the WTO in 1994, the TRIPS agreement requires members to pursue some form of intellectual property rights enactment within their own country. As Jack Ralph Kloppenburg points out in *First the Seed*, "the enactment of PBR (Plant Breeders' Rights) legislation becomes the platform and justification for the deemphasis of public breeding programs as well as the precursor for the eventual introduction of patents."

and the seed becomes an instrument of poverty and underdevelopment, one that has displaced huge numbers of farmers.

A new paradigm for seed must be based on holistic principles that consider the health of the land and those who benefit from it rather than a one-dimensional concentration on crop yields. It is imperative that all forward-looking agricultural systems first consider changing climatic conditions and be designed to reduce the waste of energy and natural resources. As an alternative to the present system that increasingly stresses "intellectual property rights," the new seed paradigm will be acted out on the local level by citizens cognizant of the complexity of natural systems.

Diversity is the ultimate form of food security; a holistic agricultural system, innovative by definition, includes a diversity of cultures and a diversity of producer-consumer relationships. "Freedom of Seed" proclaims the right of farmers to save seeds and to breed new varieties, to exchange and trade seeds, to have access to open source seed (seed free of patents), and to be protected from contamination by GMO crops.

Who will listen, one wonders, or act to change a system that is injurious to itself? Vandana Shiva, a tireless traveler and spokesperson for the sanctity of the commons, has an answer that draws on her native country's legacy of activism. At an international biotech meeting in 1987, an industry executive had claimed that five corporations would soon control the world's health and food networks, and a journalist then asked

Shiva how she would "respond to this kind of power and influence." Shiva replied:

> There was another moment in history when 85 percent of the planet was controlled by one island nation. When an old man pulled out a spinning wheel, people said, how can you defeat the British Empire with this spinning wheel? And he said, it's precisely because it looks so small, because it can be in the hands of everyone, that it is a powerful instrument.

The seed Manifesto on the Future of Seed closes with an invitation: "Living alternatives, seeds of hope." It is obvious to those who participate in living alternatives just what this means: Hope is encased in seeds, powerful in rest, in the form of potential, or possibility. I like the metaphor offered by Lu Hsün, used as an epigram to "Hope's Edge" by Frances Moore Lappé and Anna Lappé:

> Hope cannot be said to exist, nor can it be said not to exist.
> It is just like the roads across the earth.
> For actually there were no roads to begin with,
> But when many people pass one way a road is made.

The values set down at Terre Madre are put into practice at Navdanya, located on the border of Tibet and Nepal, a conservation foundation whose mission is to support local farmers and to preserve vanishing crops. Navdanya operates an organic farm on 45 acres and is actively conserving more than 5,000 crop varieties in a seed bank in Uttarakhand, north India (including 3,000 varieties of rice, 150 of wheat, and 150 of kidney beans). Bija Vidyapeeth (or Earth University), an educational initiative founded in cooperation with the United Kingdom's Schumacher College, offers courses allied with the Navdanya farm, based on the principles outlined in the manifesto. In his speech on the future of food, delivered at Georgetown University in 2011, Prince Charles invoked Gandhi's words to express ideas that resonate with the teachings of Navdanya: " . . . we may utilize the gifts of Nature just as we choose, but in her books the debts are always equal to the credits."

❦

AS A PLOWMAN, I know what fertile earth, lively earth smells like, just as the character imagined by Jean Giono who was called out into the night on the Gremone Plateau. For Jourdan, the stars are alive on that night, and Orion resembles the delicate flowers of Queen Anne's lace (Apiaceae family, also known as Umbelliferae, ancestor of our common carrot). He imagines the liveliness of seeds and he is receptive to the "lyricism of man's hope." On this fictional plateau, "the soil is saturated with cosmic juices" (in Henry Miller's words). The soil that I have come to know is a seedbed of this pulsing energy, a vital living resource that constantly asks the agriculturalist to adapt. As we will have to do—receptive to the lyricism that has inspired generations—in order to conserve diversity, guarantee freedom of choice for those who cultivate and save seed, and to revive the creative dialogue between earth's human and ecological communities.

CHAPTER 13

In Search of a Story

EACH SPRING, AS THE COVER crops in our fields begin to take on a deep green, I kneel and reach into the earth with my hands in search of a telltale sign of health that lodges in the roots of leguminous plants. Peas and beans (among other species, including peanuts) belong to the family Leguminosae, which ranks second to grains as our most valuable food source.

Because legumes have the ability to "fix" nitrogen in the soil, they are equally valuable as a soil amendment. We sow these amendments, or cover crops, beginning in late August each year, following an earlier crop of lettuce, say—or spinach, or radish—and continue to sow a changing variety of cover crop seed into November (particularly now that the autumns are progressively warmer). A cover crop's purpose is twofold: as a defense against erosion, either by wind or water, and to invigorate the soil with live organic matter. To those I would add a third reason: to create beauty. For beauty is more than an abstract quality—it connects the sower with the soil. When the seed first germinates, in late summer or early autumn, the green blanket that covers the field is inspiring, but in spring when the green returns, it is pure visual ecstasy.

The sign of soil health I look for takes the form of small white spherical galls (also referred to as nodules) attached to the roots of the legume. Over the years, we have experimented with a variety of legumes as cover crop seed—field peas, vetch, bell beans, and my current favorite, 'Austrian Winter' peas. Two thousand years ago, the Romans observed that the use of legumes could restore fertility in a field, and Virgil refers to the unusual appearance of the plant roots of this family. It was not until 1837 that a French chemist concluded that clover plants were somehow able to collect nitrogen from the air and to "fix" it in the earth.[1] Researchers from multiple nations studied the

---

1   Our atmosphere is composed of 78 percent nitrogen, and the air over each acre of the earth contains 36,000 tons of nitrogen, though this abundance is not freely available to plants. The symbiotic relationship between the legume plant and soil bacterium facilitates the creative exchange between soil and air.

process, and in 1889, a Dutchman identified the responsible organism—not the plant itself, but the millions of bacteria alive in each nodule. Nitrogen is the most difficult element for living matter to absorb, thus the job performed by bacteria is invaluable. Legume bacteria are responsible for most of the nitrogen that collects in the soil, though free-living bacteria were likely among the first inhabitants of the land, when the oceans originally receded. A buildup of nitrogen in the soil was a requirement for other life forms to come into being.

The relationship between the legume plant and the bacteria in the soil is what matters, and each bacterium is attracted to a particular variety of plant. Through its roots, the growing plant provides carbohydrates and energy in the form of sugar for the bacteria, and the bacteria collect nitrogen from the air to provide for the growing plant. When I discover the roots of 'Austrian Winter' peas covered with white nodules, I am witness to this friendship and am reassured that our fields have gained substantial fertility.

For a farmer, the practice of sowing and tending to grains and legumes is so satisfying and rewarding, I have often remarked (while leading a walkabout) that I would be content to be a grower of cover crops. Each returning year, however, our farm members and our wholesale customers continue to request a weekly supply of tomatoes, potatoes, lettuce, root crops, and squash. My salary is reliant on vegetables with abundant flavor that also look good on a plate, but the fertility of our fields—and ultimately the quality of the food—is the result of the autumn sowings of oats and bell beans, 'Austrian Winter' peas and rye.

❦

THE THEOLOGIAN AND cultural historian Thomas Berry says
that as a culture, "We are in trouble because we do not have a
good story." We are between stories, he says, and the old story
doesn't work anymore. I believe him when I observe the young
people I work with in the fields—disenchanted with the legacy
of industrial farming and the relentless pollution of our environment, they have chosen to help build the foundations for an
ecological conscience. To invent a new story, Berry counsels,
we have to recognize the "mystique of the land" to counter the
industrial mystique, and he emphasizes the need for three
commitments: a commitment to the earth; a commitment to
what he calls the ecological age—where values that support the
environment are ascendant, and we recognize our role as stewards and participants; and a commitment to a definition of
progress that includes the natural as well as the human world.

The industrial mystique is such an obvious force—we are
surrounded by the sight and sound of it—but if you need a
clearer picture of the mystique of the land, travel to a place
beside a river adorned with alders or willows, walk through a
stand of spruce or pine, sit for a moment in the silken down
under a cottonwood tree, or follow the path of a monarch
through a patch of self-seeded milkweed. In early spring, pause
under a group of oaks to witness a cloud of pollen grains disperse into the air; later, in summer, stand under a maple to
watch the winged seeds, samaras—that resemble "the innermost wings of grasshoppers" (from Gerard's *Herball* of 1597)—

spin out on the wind in search of a place to germinate. Find some common clover in flower, and look closely to witness the variety of pollinators—bees, moths, flies, and wasps—that travel among the angiosperms, ensuring our food supply. Listen as water filters over stone in a creek bed or, at the base of a waterfall, follow the spray that leads up into the air where a hawk may hover or a kingfisher may glide before slipping into the spruce woods. The land speaks to us through the creatures and plants that inhabit a place, and the stories that germinate like seeds hold a power to sustain our interconnection with other species and life-forms.

For more than 30 years, I have been tending the land, and I remain conscious of all that the word *husbandry* implies: "to manage thriftily, to economize . . . to cultivate" (words that are also essential for any poet plying his or her craft). This also implies doing so in a sustainable way. But what does it really mean to cultivate land using sustainable practices? Wendell Berry, writing in the *Temenos Academy Review*, sees a connection between the fertility cycle, or the building of soil health, and the broader cultural cycle, or the building of community health:

> The problem of sustainability is simple enough to state. It requires that the fertility cycle of birth, growth, maturity, death, and decay—what Albert Howard called "the Wheel of Life"—must turn continuously in place, so that the law of return is kept and nothing is wasted. For this to happen in the

stewardship of humans, there must be a cultural cycle, in harmony with the fertility cycle, also continuously turning in place. The cultural cycle is an unending conversation between old people and young people, assuring the survival of local memory, which has, as long as it remains local, the greatest practical urgency and value. This is what is meant, and is all that can be meant, by 'sustainability.' The fertility cycle turns by the law of nature. The cultural cycle turns on affection.

The primary motive for good care and good use is always going to be affection, because affection involves us entirely.[2]

Across this country—and, indeed, throughout the world—in the realm of agriculture and seed preservation, there are multiple examples of people working together to create a new story. Collaboration is key to ensuring that the revival of interest we are experiencing now continues to build and to become a "vibrant web of biodiversity and resilience," in the words of seedsman Bill McDorman. When our community farm participated in the Public Seed Initiative and the Organic Seed Partnership, two programs that facilitated on-farm

---

2   To link this chapter with the previous one, I will add an observation by Rebecca Solnit, the author of *The Faraway Nearby* and other books: " . . . *free* has the same Indo-European root as the Sanskrit word *priya*, which means 'beloved' or 'dear.' If you think of etymology as a family tree, the dictionary says that most descendants of that ancient ancestor describe affection, and only the Germanic and Celtic branches describe liberty."

collaboration among multiple participants—the Northeast Organic Farming Association of New York; our land grant university, Cornell; the USDA Agricultural Research Service in Geneva, New York; and farmers—the communication lines opened up and valuable partnerships were fostered. Fedco Seeds founder C. R. Lawn points out that preservation of heritage varieties is only the beginning of the seed revival; for thousands of years, farmers have been the ones responsible for observing plants in the field, for noticing favorable mutations, for selecting and for stabilizing a variety. Much of that basic knowledge has been replaced, or lost, over the last century, so farmer-to-farmer communication and farmer-breeder collaborations are essential to expanding our agricultural diversity. The industrial agricultural paradigm dominant at the start of the 21st century is only 60 years old—Lawn notes that "There were no seed companies until the 18th century . . . no university breeding programs until the 19th . . . no hybrids or multinational seed conglomerates until the 20th"—so there is no reason to assume that we are stuck with it.

Bill McDorman, who founded a number of seed companies beginning in the early 1980s, is encouraged by the renaissance in independent seed companies over the last 30 years, asserting: "We are returning full circle to the regionalism, resilience, and genetic abundance at the dawn of the last century." Despite his long association with seed companies, and his current position as executive director of Native Seeds/ SEARCH (a nonprofit founded by Gary Paul Nabhan, among others), McDorman impishly remarks that he would prefer to put seed companies and seed banks out of business. Saving seed does not have to be complicated, he insists, and he hopes to inspire wide communities of growers and seed savers. Located in Tucson, Arizona, Native Seeds/SEARCH operates its Seed School multiple times during the year, graduating "seed citizens" who will carry on the practice in their own gardens and give practical guidance to others on how to preserve biodiversity.

John Navazio, plant breeding and seed specialist for Washington State University Extension and the cofounder and senior scientist for the Organic Seed Alliance, observes that once you develop an interest in seed, you are not involved in just another project but in a process. Farmers once participated in the entire life cycle of the plants they cultivated, and Navazio is working with growers to once again embrace the whole natural cycle through seed saving and plant breeding. The word *heirloom* now sparks considerable interest,[3] though, as a breeder, Navazio focuses on the "heirlooms of tomorrow" and on teaching farmers to continuously select for traits in OP (open-pollinated) varieties that perform well in a changing climate.

The focus of Navazio's work is to ensure that genetic variability continues to circulate, so he is always selecting for what he calls "good genetic breadth"—precisely what is missing

---

3   Recently, shopping at our stand at a local farmers' market, one woman turned to her friend and said, sounding impressed, "They sell heirloom lettuce at this stand!" The reply (imagine a significant Long Island accent): "What d'ya mean, it grows in the air?"

in hybrid systems. And through what is known as "participatory plant breeding" (PPB), he is passing this on to the next generation of seed stewards. The Organic Seed Alliance has created a Farmer Seed Stewardship network with a mission to conserve, improve, and develop seeds that are adapted to regional needs.

Tom Stearns, founder of the High Mowing Organic Seeds company, sent out his first seed listing in 1996 when he was age 20—I still have a photocopy of his pamphlet describing 28 vegetable varieties, from which we purchased a few packets to support the cause. The High Mowing catalog is now 64 pages and lists more than 600 varieties of certified organic seed—extremely rare for a seed catalog. High Mowing trials many of their own varieties in demonstration fields in Wolcott, Vermont, though they also contract with other farmers to grow seed using "stock seed" (seed first grown in the High Mowing fields). "This is a very important thing," Stearns says, "because this is the first step in quality control."

High Mowing now operates its own breeding program, initially funded through investments based on the Slow Money model. In 2010, the company released two new OP varieties: 'King Crimson' pepper, and a beautiful dark zucchini, 'Midnight Lightning'. Stearns maintains that organic farmers do not have "the right seed genetics to support true organic farming methods," so he is working to change that. Recently, High Mowing hosted an Organic Seed and Breeding Field School, which offered hands-on participatory workshops for students of ecological agriculture. Though information relating to breeding and seed saving has been closely guarded by private companies during the last century, High Mowing is committed to moving toward the shared goal of self-sufficient, healthy food communities. Stearns asserts: "We need as many people as possible to know how to save seeds so that the interdependence of our food system is strengthened."

Nathan Corymb, who has worked with community agriculture and for Turtle Tree Seed for many years, recently presented the outline of an idea to stimulate community-supported seed production, exchange, and distribution at a national biodynamic agriculture conference. Just as people have become divorced from the source of their food, so farmers and gardeners have become divorced from the source of our seed. It was the Biodynamic Farming and Gardening Association that sponsored the first meetings of CSA farmers and supporters in a Waldorf school in Kimberton, Pennsylvania, and Corymb hopes to build on a model that has grown to 6,000 or more farms throughout the country.

In 2010, at an event I never could have imagined—an heirloom vegetable auction at Sotheby's, New York, for which our 'Amagansett Peach' garlic had been chosen—I happened to meet another pioneer of the seed business, Jere Gettle. He mailed his first seed listing at age 17, with seeds saved from his gardens near Baker Creek, in the Ozark range of southern Missouri. Gettle went on the road a few years later, to Mexico and then Thailand, in search of seeds. In the winter of 2000, I was looking for a rare varietal I admired but could not source— Turkish eggplant (a round orange fruit, easily mistaken for a tomato)—and a friend directed me to Baker Creek Heirloom Seeds. Since then, our plant selection has been augmented by such varieties as 'Pipian from Tuxpan' (a beautiful and rather tasteless striped squash, though when roasted its seeds are delicious), 'Malakhitovaya Shkatulka' ("malachite box," a graceful green tomato from Russia), and the huge, candy-flavored watermelon 'Crimson Sweet'.

The Baker Creek catalog, now 212 pages long, lists more than 1,400 varieties of (primarily) heirloom seeds. Gettle is an inveterate traveler—passing through the small town of Petaluma, California, with his wife, Emilee, a few years ago, he happened to notice a bank building "for lease." Soon the sign on the facade of the classic western mercantile building had been changed to read: SEED BANK. The currency offered within is not greenbacks but OP heirloom seeds that will preserve a green diversity. On a trip to the East Coast, when Gettle learned that the oldest continuously operating seed company in the United States was up for sale, he had to inquire. Comstock, Ferre & Co. was established in 1811; throughout the 19th century, seed purveyors set off by wagon from Wethersfield, Connecticut, to supply settlers in the near "West" with seeds for the garden and farm. Baker Creek Heirloom Seeds purchased the company, and their seed packets are now for sale in the historic Comstock barn. The old building reverberates with music during an annual summer festival that honors the rich New England agricultural tradition (beekeepers, basket weavers, candle makers, storytellers) and educates visitors about the risks of genetic engineering.

Baker Creek Heirloom Seeds launched its National Heirloom Exposition in Santa Rosa, California, in 2011. The following year, more than 10,000 people attended the 3-day exposition, including several representatives from the growing number of US seed libraries. A seed library works on the same principle as a book-lending library—seeds are distributed to gardeners, planted out in a home garden, a few plants are left to go to seed, and the gardener returns the saved seed to restock the lending library. Ken Greene, cofounder of the Hudson Valley Seed Library, was one of those present who joined with other "seed librarians" to call for the formation of a national association.

When Greene was working in the Gardiner Library in upstate New York, he happened to meet Sascha DuBrul, the founder of the Bay Area Seed Interchange Library (BASIL), the first seed-lending library in the country, and he was intrigued by the connection between culture and agriculture, books and seeds. Soon the Gardiner Library was also lending seeds, and Greene was learning and teaching the practice of seed saving. At the same time, Rebecca Newburn, a middle school science teacher, persuaded the main branch of California's Richmond Public Library to establish Richmond Grows, a lending program, and to start a demonstration garden on the library grounds.

In 2008, Greene and his partner bought a few acres of land in Accord, New York, and expanded their educational outreach to engage the local community; the seed library is now organized online, they communicate their ideals through timely blogs, and membership in the organization is doubling each year. Greene is aghast that no one demands accountability or transparency of the large seed companies. Those working to preserve seed heritage should be required to make ethical choices, he says, and he acknowledges the common sense expressed by the Oregon seedsman Frank Morton: "Make your own selections, and be sure that everyone has access to them!"

Morton, founder of Wild Garden Seed, who has inspired a generation of seed lovers and plant breeders through his innovative techniques and his creative garden aesthetics, offers a definition of a seed that could well serve as the epigraph for the new story we are searching for: "The best deal in nature: dense nutritional matter with a self-organizing program and energy array. For cheap."

A seed is a living thing that embodies roots, stem, leaves, and fruit in an embryonic state and retains the ability to convert the sun's energy into a source of food. Over the last 10,000 years, our ancestors learned to identify and select for desired traits—we should acknowledge that the seeds of our daily bread were the gift of diverse wild habitats and generations of attentive stewardship.

♃

DURING THE COURSE of writing this book, I have been inspired by numerous texts and illustrations, but two have remained on my desk throughout. One is a photograph that appeared on the front page of the *New York Times* at last year's spring equinox under the headline "The Cosmos, Back in the Day," depicting an image from data recorded by a European Space Agency satellite. This heat map of the universe as it appeared 370,000 years after the Big Bang is oval shaped and pulses with blue and red colors against the black of space. In its infancy, at least in this human recording of it, the universe resembled a seed.

The other is a very small book, *The Secret Life of the Flowers*, by Anne Ophelia Dowden, one in a series of titles from the Odyssey Library. Under an illustration of the anatomy of a flower is this description: "Pollen grain germinates on mature stigma, sends tube down into ovary to fertilize ovule." As I contemplate the abstract image of our universe as it appeared 13 billion years ago, I also want to acknowledge the beauty and precision of the plant structure that evolved to produce seeds.

Earlier, I referred to Thomas Berry's exhortation that we commit to a definition of progress that includes the natural as well as the human world. The author Ronald Wright uses the term "progress trap" to describe a short-term social or technological improvement that turns out, in the long term, to be a backward step—once the mistake is realized, it is too late to change course. In the modern industrial era, our human species seems to be more than capable of taking backward steps. But there is another way to proceed, through a recognition of the web of relationships of which we are a part. In *The Myth of Progress*, my friend Tom Wessels offers the example of mychorrizal fungi—"fungus-roots" that help plant life absorb nutrients—in a certain forest system. The fungi actually spread out within the soil to feed energy to struggling paper birch trees when they are surrounded by more robust Douglas fir trees. He calls them "tree shepherds." He adds that our task as individuals is to progress in a manner in which our attention, our compassion, and our empathy grow ever outward to benefit our communities, and society as a whole.

In haunting words that I return to often, at the close of his book-length essay *The Tree,* the English novelist John Fowles speaks of a namelessness that "is beyond our science and our arts because its secret is being, not saying." He writes:

> We still have this to learn: the inalienable otherness
> of each, human and non-human, which may seem
> the prison of each, but is at heart, in the deepest of
> those countless million metaphorical trees for which
> we cannot see the wood, both the justification and
> the redemption.

It is that namelessness, whose secret is being, that is the life force embedded in an embryo, poised for germination, at the heart of every seed. There is a *fabric of relationships* available to us encoded in seeds, part of a timeless refrain that invites us to participate, to renew the story that has shaped us.

❦

AT THE END of this journey into seedtime, I step out into the scrub oak woods that surround my house and breathe in the moisture that arrives from the ocean at Sagaponack, a few miles to the south. I stand under a handsome 40-foot-tall black pine that fans out over the oaks and, though I have known this tree for 20 years, for the first time I notice the fluid turns of its body—evidence of growing toward and against the force of hurricane winds that sweep across this island—as sunlight filters through to illuminate the rough reddish bark. Seeds cover the earth we stand on, the tree and me—male and female cones, so different in shape, spread out over a thick layer of long needles, the carbon cover over the sandy soil, soon to be humus.

The cicadas have just begun to sing, and though unseen, they are part of the *miracle of the realness* revealed in this wood. Their tentative notes will soon swell to a symphony that will dominate the August nights. Above the needles, minerals, and flakes of bark, oak and hickory leaves create a harmony with the west wind, and I am nourished by the silence beyond all thought. Listening, I accept the friendship of the seasons, and I offer it to you.

AFTERWORD

*"To what useful end could I use my eyes without acknowledging that
they are only one of the earth's inexhaustible ways of seeing?"*

John Hay, *The Immortal Wilderness*

AN OUTCROP OF GRANITE ROCKS lying just off the coastline of
Cornwall's Penwith Peninsula shelters the small harbor of
Mousehole village, where I lived with my wife and our first son
during the decade of the 1980s. The village was once named
Porth Enys, the island port, for this rugged outcrop, and from
our Love Lane cottage, on the hillside above the village, we
also looked out at Merlin's rock, prominent at low tide and in
calm weather. The lines entitled "Envoi," which I wrote to
introduce a book of poetry, evoke the seascape:

First light rides rocking water,
    stone blooms brush by windrush
        on sky and the Crackers.
Herbs and bone illumine dark earth
    where blue stars sleep
        under the candles of alchemilla.

I learned to turn the earth and to cultivate crops in the dark, fertile soil of the cliff meadows perched just above Merlin's rock, where the interconnection between mineral, flowering plants, and man—far from an abstraction—could be immediately felt. Though these meadows had largely been abandoned by market gardeners, their careful stewardship, carried on for generations, was evident in the artful stone walls used for terracing, the hedges of privet and fuchsia planted to block the strong winds, and the thousands of daffodils and narcissi that blossomed along the pathways each spring.

This landscape, home to a colorful diversity of plants, is source material for some rich Celtic mythology, and the stories are woven into the greenery and exquisite flowers of the hedges, and the sea foam. One tale explains the disappearance of master magician Merlin: After instructing his lover, the beautiful Niniane, in the deepest secrets of his art, he falls asleep in her arms under a hedge. She unwinds her long veil to cover Merlin and the whitethorn bush and circles them nine times, whispering the words the magician has taught her. The spell proves to be unbreakable—Niniane is free to come and go, but Merlin is forever captive in the whitethorn. Though he has given over his power, the wizard is at home in timelessness—and, as noted by the storyteller, Heinrich Zimmer, "the whitethorn bush blossoms imperishably." The unity of the human and natural realms is revealed year after year when the stamen and pistil of the whitethorn open to the air on the granite headland.

It was in this storied landscape that I learned to respect the resiliency of soil and plants and seeds as part of an evolving, cyclical process that is nourished daily by the energy of the sun. Each seed, I came to see—manifest in a multitude of shapes and sizes—stores that energy in the invisible form of leaves and branches, or petals and fruit. And in a place characterized by mild temperature and abundant rainfall, each seed, planted into well-nurtured soil, enthusiastically revealed the fecundity of nature. Five feet above the ground, valerian and campion sprouted from the stone hedges (granite rocks bound together with soil), wild garlic and nettles flourished along the lanes, and fuchsia flowered wherever a branch was planted in the earth. At home here (because of my Celtic roots and the artful, expressive voice of the land), I embraced my role as steward, and acknowledged kinship with the German poet-philosopher Goethe, who labored to understand plants and wrote: "The simpler powers that lie deep within nature" can be revealed "as purely as the objects of the visible world are formed in a clear eye."

I feel fortunate to have found my way from the infancy of suburbia, on the outskirts of Buffalo, New York, to the cliffside meadows of Cornwall, and then to the tip of another peninsula also touched by the tidal surges of the Atlantic—Long Island—where the book of nature is less obscured, more readable. After a year in Oxford, England, mostly spent reading literature (in bliss) in the historic Bodleian, I stepped out of the library into the garden, and then into larger fields, to learn something of the diverse community—from microbial life to flowering trees—that provide our food. The practice of agriculture, like other human industries, can severely obscure the landscape too, but if performed with a seasoned restraint and the intention to protect diversity and to improve soil fertility, a farmer can raise a crop and also enhance the delicate balance between the cultivated and the wild. "Poetry and farming are surprisingly similar art forms," my friend who farms with her family at Mountain Dell Farm, Lisa Wujnovich, writes. "In farming I toil at what nature does easily, and in poetry I work at the ephemeral by attempting to embody time in language."

Seeds, like words, "behave like capricious and autonomous beings," so if we give them space to perform, perhaps we stand a chance to inherit their intelligence. This seems like a wiser choice to me, rather than to force our intelligence upon them.

I began this book with an observation made by Cary Fowler more than 20 years ago: "We are in the midst of a mass extinction event in agriculture . . . " and the trend that promotes monocultures and economic consolidation has only intensified in the ensuing years. In 1985, in his book, *Altered Harvest: Agriculture, Genetics, and the Fate of the World's Food Supply,* Jack Doyle wrote:

> The genetic centralization of food production *is* something to worry about. The stuff of national sustenance is involved: the sustaining, essential

ingredients of life; the most basic components of food-making. These are the food determinants at the innermost sanctum of biology. Indeed, these are the ingredients of power; these genes that command the faculty of chloroplasts to bottle the sun so we can have food. They are, in their workings, the closest of a kind to anything we dare call sacrosanct. We owe them more respect than to patent their trade.

The health of our fields, the health of our plant communities, and the future of our food supply will depend on whether, as a global culture, we can learn to respect the whole of the biological community, and to accept our role as citizens of it (and to honor those who still retain the connection).

I like the example of clover. I went out today to inspect the autumn crops, those we will harvest into December—cauliflower, broccoli, kale, and collards—and those we will harvest for the Winter Share, to be stored in the root cellar—turnips, radish, rutabaga, and carrots. In the adjacent field, where we harvested wheat in July, the flowers of red clover dominate the green blanket that covers the silt loam. After the wheat harvest, I mowed the chaff that remained, and the clover, sown 2 years ago and at home in the shadow of the wheat, was stimulated by this increased access to the light. It is a beautiful plant—three heart-shaped leaves delicately streaked with a whitish V—and at the end of an 18-inch stem, hundreds of purple-red flower whorls holding the space for a seed. It is the tiniest seed we plant, hardly visible in the center of your palm, yet it is responsible for the most significant contribution to field fertility.

Having germinated in the shade of the wheat, the delicate white flowers of campion—otherwise uncommon in our landscape— complete the natural bouquet of the field.

Our culture, our habitation in this time on Earth, is in need of transformation, some say in the shape of a new story. Transformation comes from within, and seeds have mastered the art. *Clover does not think about responsibility . . . its response to altered circumstances is to give nourishment.* I have sown the seeds of clover and seen it come to flower in the fields, and I have tasted the harvest of nutritious food the following season. May we continue to cultivate our fields with the imperishable mystery in mind and to playfully, carefully follow these seeds and nurture them, where they desire to go.

SOURCES

Abram, David. *The Spell of the Sensuous.* New York: Vintage Books, 1997.

Adelson, Glenn, James Engell, Brent Ranalli, and K. P. Van Anglen, eds. *Environment: An Interdisciplinary Anthology.* New Haven, CT: Yale University Press, 2008.

Aitken, Robert. *The Mind of Clover: Essays in Zen Buddhist Ethics.* Berkeley, CA: North Point Press, 1984.

Altieri, Miguel A. *Genetic Engineering in Agriculture: The Myths, Environmental Risks, and Alternatives.* Oakland, CA: Food First Books, 2004.

Anderson, Ross. "After 4 Years: Checking Up on the Svalbard Global Seed Vault." *Atlantic,* February 28, 2012.

Antoniou, Michael, Claire Robinson, and John Fagan. *GMO Myths and Truths.* Earth Open Source, 2012.

Armstrong, Karen. *A Short History of Myth.* Edinburgh: Canongate, 2005.

Ashworth, Suzanne. *Seed to Seed: Seed Saving and Growing Techniques for Vegetable Gardeners.* Decorah, IA: Seed Savers Exchange, 2002.

Attenborough, David. *Life on Earth.* Little, Brown and Company, 1979.

Barks, Coleman, trans. *Rumi: Bridge to the Soul: Journeys into the Music and Silence of the Heart.* New York: Harper One, 2007.

———. *The Glance: Songs of Soul-Meeting.* New York: Penguin Compass, 1999.

Berry, Thomas. *Creative Energy: Bearing Witness for the Earth.* San Francisco: Sierra Club Books, 1988.

———. *The Great Work: Our Way into the Future.* New York: Three Rivers Press, 1999.

Berry, Wendell. *The Unsettling of America: Culture and Agriculture.* San Francisco: Sierra Club Books, 1996.

———. *Bringing It to the Table: On Farming and Food.* Berkeley, CA: Counterpoint Press, 2009.

———. "The Agrarian Standard." *Orion* (Summer 2002).

Bioneers. *Visions for a 21st Century Agriculture.* Santa Fe: Bioneers/Collective Heritage Institute, 2003.

Boutard, Carol, and Anthony Boutard. "Ancient Crop Wins New Following." *Growing for Market* 15, no. 7.

Bowling, Maria. www.mariabowling.com.

Brown, Lauren. *Grasses: An Identification Guide.* Houghton Mifflin, 1979.

Buhner, Stephen Harrod. *The Secret Teachings of Plants, in the Direct Perception of Nature.* Rochester, VT: Bear & Company, 2004.

Bunting, Basil. *Collected Poems.* Moyer Bell, 1985.

————. *Complete Poems.* New York: New Directions, 2000.

Caduto, Michael J., and Joseph Bruchac. *Native American Gardening: Stories, Projects and Recipes for Families.* Golden, CO: Fulcrum, 1996.

Campbell, Joseph. *The Mythic Image.* Princeton, NJ: Princeton University Press, Bollingen Series C, 1974.

Capon, Brian. *Botany for Gardeners: An Introduction and Guide.* Portland, OR: Timber Press, 1990.

Carson, Rachel. *Silent Spring.* New York: Houghton Mifflin, 1962.

————. *Under the Sea-Wind.* New York: Simon and Schuster, 1941.

Chaskey, Scott. *This Common Ground: Seasons on an Organic Farm.* New York: Viking, 2005.

————. *Cardine: A Book of Odes.* Waterhouse's Clogh, 1980.

————. *Stars Are Suns.* York, UK: Stoneman Press, 1995.

Connolly, Bryan, and Rowen White. *Breeding Organic Vegetables: A Step-by-Step Guide for Growers.* Rochester, NY: NOFA-NY, 2011.

————. *The Wisdom of Plant Heritage: Organic Seed Production and Saving.* Rochester, NY: NOFA Interstate Council, 2004.

Cutler, Karan Davis. *Starting from Seed: The Natural Gardener's Guide to Propagating Plants.* Brooklyn, NY: Brooklyn Botanic Garden, 1998.

Darwin, Charles. *On the Origin of Species.* New York: The New American Library, 1958 (first published by John Murray, 1859).

Davenport, Guy. *Every Force Evolves a Form.* Berkeley, CA: North Point Press, 1987.

Deakin, Roger. *Wildwood: A Journey through Trees.* New York: Free Press, 2007.

Deppe, Carol. *Breed Your Own Vegetable Varieties: The Gardener's and Farmer's Guide to Plant Breeding and Seed Saving.* White River Junction, VT: Chelsea Green, 2000.

Diamond, Jared. *Guns, Germs, and Steel: The Fates of Human Societies.* New York: W. W. Norton, 1997.

Dowden, Anne Ophelia. *The Secret Life of the Flowers.* New York: Odyssey Press, 1964.

———. *From Fruit to Flower.* New York: Ticknor and Fields, 1994.

Doyle, Jack. *Altered Harvest: Agriculture, Genetics, and the Fate of the World's Food Supply.* New York: Viking Penguin, 1985.

Dunne, Claire. *Wounded Healer of the Soul.* New York: Parabola Books, 2000.

Eiseley, Loren. *The Star Thrower.* New York: Harcourt, Brace, Jovanovich, 1978.

Eliot, T. S. *The Complete Poems and Plays.* New York: Harcourt, Brace, and World, 1971.

Engeland, Ron. *Growing Great Garlic.* Okanogan, WA: Filafree Productions, 1991.

Erdoes, Richard, and Alfonso Ortiz, eds. *American Indian Myths and Legends.* New York: Pantheon, 1984.

Farb, Peter. *Living Earth: The Story of the Marvelous Abundance and Complexity of Life within the Soil beneath Us.* New York: Harper & Brothers, 1959.

Forbes, Peter, Ann Armbrecht Forbes, and Helen Whybrow, eds. *Our Land, Ourselves, Readings on People and Place.* San Francisco: The Trust for Public Land, 1999.

Forbes, Peter. *The River.* Waitsfield, VT: Sabbath Year Series, 2012.

Fowler, Cary, and Pat Mooney. *Shattering: Food, Politics, and the Loss of Genetic Diversity.* Tucson: University of Arizona Press, 1990.

Fowles, John. *The Tree.* New York: Ecco/Harper Collins, 2010.

Frazer, Sir James George. *The Golden Bough: A Study in Magic and Religion.* New York: MacMillan, 1951.

Fukuoka, Masanobu. *The One-Straw Revolution.* Emmaus, PA: Rodale, 1978.

Gallagher, Tess. *Amplitude.* Minneapolis: Graywolf Press, 1987.

Gettle, Jere, and Emilee Jere. *The Heirloom Life Gardener.* New York: Hyperion, 2011.

Gibson, Margaret. *Earth Elegy: New and Selected Poems.* Baton Rouge: Louisiana State University Press, 1997.

Hart, Kathleen. *Eating in the Dark: America's Experiment with Genetically Engineered Food.* New York: Pantheon, 2002.

Hawken, Paul. *Blessed Unrest: How the Largest Movement in the World Came into Being, and Why No One Saw It Coming.* New York: Viking, 2007.

Hay, John. *The Immortal Wilderness*. New York: W. W. Norton, 1987.

Heaney, Seamus. *Field Work*. New York: Farrar, Straus, Giroux, 1976.

Heinberg, Richard. "Fifty Million Farmers." *Annals of Earth* 26, no. 1.

Hobhouse, Henry. *Seeds of Change: Five Plants That Transformed Mankind*. New York: Harper and Row, 1985.

————. *Seeds of Wealth: Four Plants That Made Men Rich*. Shoemaker and Hoard, 2003.

Holdrege, Craig. *Genetics and the Manipulation of Life: The Forgotten Factor of Context*. Aurora, CO: Lindisfarne Press, 1996.

Hoodes, Liana, and Michael Sligh, eds. *National Organic Action Plan, Advancing Organic Agriculture in the U.S.* Pittsboro, NC: Rural Advancement Foundation International-USA, 2010.

Howard, Sir Albert. *An Agricultural Testament*. Oxford University Press, 1940.

————. *The Soil and Health*. Schocken Books, 1947.

HRH The Prince of Wales. *On the Future of Food*. Emmaus, PA: Rodale, 2012.

Jackson, Wes. *Call for a Revolution in Agriculture: First Annual E. F. Schumacher Lectures*. Bristol, UK: E. F. Schumacher Society, 1981.

Kesseler, Rob, and Wolfgang Stuppy. *Seeds: Time Capsules of Life*. Earth Aware Editions, 2012.

Kessler, Milton. *Free Concert*. Wilkes-Barre, PA: Etruscan Press, 2002.

Kimbrell, Andrew. *The Fatal Harvest Reader: The Tragedy of Industrial Agriculture*. Washington, DC: Island Press, 2002.

Kirschenmann, Frederick L. *Cultivating an Ecological Conscience: Essays from a Farmer Philosopher*. Lexington, KY: University Press of Kentucky, 2010.

Kittredge, Jack, and Julie Rawson, eds. "Special Supplement on Transgenic Crops." *Natural Farmer* (Spring 2012).

Klindienst, Patricia. *The Earth Knows My Name*. Boston: Beacon Press, 2006.

Kloppenburg, Jack Ralph Jr. *First the Seed: The Political Economy of Plant Biotechnology, 1492–2000*. Madison, WI: University of Wisconsin Press, 2004.

Kunitz, Stanley. *Passing Through: The Later Poems*. New York: W.W. Norton, 1995.

Lappe, Frances Moore, and Anna Lappe. *Hope's Edge: The Next Diet for a Small Planet*. New York: Jeremy P. Tarcher/Putnam, 2002.

Lawn, C. R. *Growing Habitat, Embracing Biodiversity*, 2010; *Why Save Seeds*, 2001; *Honoring Plant Breeders*, 1998; *Debunking Biotech Myths*, 2001; *Surge, Siege, and Seed*, 2013; Waterville, ME: Fedco Seeds.

Laws, Bill. *Fifty Plants That Changed the Course of History*. New York: David and Charles, 2010.

Lazor, Jack. *The Organic Grain Grower: Small-Scale, Holistic Grain Production for the Home and Market Producer.* White River Junction, VT: Chelsea Green Publishing, 2013.

Lemon Street Gallery. *Breon O'Casey, A Decade.* Cornwall, UK: 2009.

Levi-Strauss, Claude. *The Raw and the Cooked.* New York: Harper and Row, 1969.

Lord, Tony. *Flora.* London: Weidenfeld & Nicolson, 2003.

Mandelshtam, Osip. *Selected Poems.* London: Penguin, 1991.

Mandelstam, Nadezhda. *Hope against Hope.* New York: Modern Library, 1999.

Mas Masumoto, David. *Wisdom of the Last Farmer.* New York: Free Press, 2009.

Merrill, Richard, ed. *Radical Agriculture.* New York: Harper Colophon, 1976.

Mooney, Pat Roy. *Seeds of the Earth: A Public or Private Resource?* Ottawa, ON: Inter Pares, 1979.

Moore, Kathleen Dean, and Michael P. Nelson, editors. *Moral Ground.* San Antonio, TX: Trinity University Press, 2010.

———. *Riverwalking.* New York: Lyons & Burford, 1995.

Nabhan, Gary Paul. *Enduring Seeds: Native American Agriculture and Wild Plant Conservation.* Tucson: University of Arizona Press, 1989.

———, and Buchmann, Stephen L. *The Forgotten Pollinators.* Washington, DC: Island Press, 1996.

———. *Cultures of Habitat: On Nature, Culture, and Story.* Berkeley, CA: Counterpoint Press, 1997.

———. *Where Our Food Comes From: Retracing Nikolay Vavilov's Quest to End Famine.* Washington, DC: Island Press, 2009.

Navazio, John. *The Organic Seed Grower: A Farmer's Guide to Vegetable Seed Production.* White River Junction, VT: Chelsea Green, 2012.

Neruda, Pablo. *Odes to Common Things.* Boston: Bulfinch Press, 1994.

Nestle, Marion. *Food Politics.* Berkeley, CA: University of California Press, 2003.

Neumann, Erich. *The Origins and History of Consciousness.* Princeton, NJ: Princeton University Press, Bollingen Series XLII, 1970.

Niedecker, Lorine. *New Goose.* Rumor Books, 2002.

———. *T & G: The Collected Poems (1936–1966).* Jargon Society, 1969.

Olds, Jerome. *The Encyclopedia of Organic Gardening.* Emmaus, PA: Rodale, 1976.

Overbye, Dennis. "The Cosmos, Back in the Day." *New York Times,* March 22, 2013.

Paz, Octavio. *The Bow and the Lyre* (Ruth L. C. Simms, trans.). Austin, TX: University of Texas Press, 1973.

Peattie, Donald Culross. *Flowering Earth.* Bloomington, IN: Indiana University Press, 1991.

Pollan, Michael. *Second Nature: A Gardener's Education.* New York: Grove Press, 1991.

———. *The Botany of Desire: A Plant's Eye View of the World.* New York: Random House, 2001.

———. *In Defense of Food: An Eater's Manifesto.* New York: Penguin Press, 2008.

Pringle, Peter. *The Murder of Nikolai Vavilov: The Story of Stalin's Persecution of One of the Twentieth Century's Greatest Scientists.* JR Books, 2009.

Reader, John. *Potato: A History of the Propitious Esculent.* New Haven, CT: Yale University Press, 2009.

Roach, Margaret. "Look Carefully at Those Seeds." *New York Times,* March 3, 2013. http://www.nytimes.com/2013/03/03/opinion/sunday/look-carefully-at-those-seeds.html?hp&_r=0

Robin, Marie-Monique. *The World According to Monsanto: Pollution, Corruption, and the Control of Our Food Supply.* The New Press, 2010.

Robinson, Jo. "Breeding the Nutrition Out of Our Food." *New York Times,* May 26, 2013. http://www.nytimes.com/2013/05/26/opinion/sunday/breeding-the-nutrition-out-of-our-food.html?pagewanted=all

Rodale, Maria. *Organic Manifesto: How Organic Food Can Heal Our Planet, Feed the World, and Keep Us Safe.* Emmaus, PA: Rodale, 2010.

Safina, Carl. *Song for the Blue Ocean.* New York: Henry Holt and Company, 1997.

———. *The View from Lazy Point.* New York: Henry Holt and Company, 2011.

Sahn, Jennifer, ed. *Thirty-Year Plan: Thirty Writers on What We Need to Build a Better Future.* Great Barrington, MA: Orion, 2012.

Sauer, Carl O. *Land and Life.* Berkeley, CA: University of California Press, 1963.

———. *Seeds, Spades, Hearths & Herds: The Domestication of Animals and Foodstuffs.* Cambridge, MA: MIT Press, 1969.

———. *Selected Essays, 1963–1975.* Turtle Island Foundation, 1981.

Seabrook, John. "Sowing for Apocalypse: The Quest for a Global Seed Bank." *New Yorker,* August 27, 2007. http://www.newyorker.com/reporting/2007/08/27/070827fa_fact_seabrook

Shakespeare, William. *The Riverside Shakespeare.* New York: Houghton Mifflin, 1974.

Shiva, Vandana. *Stolen Harvest: The Hijacking of the Global Food Supply.* Cambridge, MA: South End Press, 2000.

———, ed. *Manifestos on the Future of Food and Seed.* Cambridge, MA: South End Press, 2007.

———. "A Worldview of Abundance." *Orion* (Summer 2000).

———. "Seed Freedom." *Resurgence,* no. 278 (May/June 2013).

———. "Everything I Need to Know I Learned in the Forest." *Yes!* (Winter 2013).

Silverstein, Alvin, and Virginia Silverstein. *Life in a Bucket of Soil.* Mineola, NY: Dover, 1972.

Snyder, Gary. *A Place in Space.* Berkeley, CA: Counterpoint Press, 1995.

Souder, William. *On a Farther Shore: The Life and Legacy of Rachel Carson.* New York: Crown, 2012.

Standage, Tom. *An Edible History of Humanity.* London: Atlantic Books, 2009.

Stewart, Keith. *It's a Long Road to a Tomato.* Marlowe and Company, 2006.

Storl, W. D. *Culture and Horticulture.* Bio-Dynamic Literature, 1979.

Thoreau, Henry D. *Faith in a Seed.* Washington, DC: Island Press, 1993.

Thuente, Joanne, ed. *Seed Savers Exchange 2013 Yearbook.* Decorah, IA: Seed Savers Exchange, 2013.

Tompkins, Peter, and Christopher Bird. *The Secret Life of Plants.* New York: Harper, 1973.

Tortorello, Michael. "Vintage Seeds or Flinty Hybrids?" *New York Times,* March 24, 2011. http://www.nytimes.com/2011/03/24/garden/24seeds.html?pagewanted=all

von Goethe, Johann Wolfgang. *The Metamorphosis of Plants.* Cambridge, MA: MIT Press, 2009.

Walke, Bernard. *Twenty Years at St. Hilary.* Anthony Mott, 1982.

Wells, Diana. *Lives of the Trees: An Uncommon History.* Chapel Hill, NC: Algonquin Books, 2010.

Wells, Spencer. *Pandora's Seed: The Unforeseen Cost of Civilization.* New York: Random House, 2010.

Wessels, Tom. *The Myth of Progress: Toward a Sustainable Future.* Burlington, VT: University of Vermont Press, 2006.

Whealy, Kent. *Garden Seed Inventory, Third Edition.* Seed Saver Publications, 1992.

Williams, Carol. *Bringing a Garden to Life.* New York: Bantam, 1998.

Williams, Terry Tempest. *Finding Beauty in a Broken World.* New York: Vintage Books, 2009.

———. *Refuge: An Unnatural History of Family and Place.* New York: Vintage Books, 1991.

———. *Red: Passion and Patience in the Desert.* New York: Vintage Books, 2002.

Wordsworth, William. *The Prelude: Or Growth of a Poet's Mind.* New York: Houghton Mifflin, 1982.

Wujnovich, Lisa. *This Place Called Us.* Stockport Flats, 2008.

Yeats, W. B. *Collected Poems.* New York: Macmillan, 1970.

Zimmer, Heinrich. *Myths and Symbols in Indian Art and Civilization.* New York: Pantheon, Bollingen Series VI, 1946.

———. *The King and the Corpse: Tales of the Soul's Conquest of Evil.* Princeton, NJ: Princeton University Press, Bollingen Series XI, 1971.

From *Acres USA, A Voice for Eco-Agriculture:*

Cesario, Cheryl. "Seeds of Success." *Acres* 41, no. 1.

McDorman, Bill, and Stephen Thomas. "Sowing Revolution: Seed Libraries Offer Hope for Freedom of Food." *Acres* 42, no. 1.

———. "Seed of Sustainability: Preserving the Past One Plant at a Time." *Acres* 41, no. 1.

Shakra, Quin. "Modern Seed Movement: Plant Breeder John Navazio Discusses the Significance of Organic Seed." *Acres* 43, no. 1.

Walters, Chris. "Right to Know: Interview with Ken Roseboro." *Acres* 41, no. 9.

From *Orion:*

Brower, Lincoln P. "Canary in the Cornfield: The Monarch and the Bt Corn Controversy." *Orion* 20, no. 2 (Spring 2001): p. 32.

Monson, Ander. "Dear Squash." *Orion* (July/August 2013).

Solnit, Rebecca. "Mysteries of Thoreau, Unsolved." *Orion* 32, no. 3 (May/June 2013): p. 18.

A FEW ADDITIONAL NOTES

A Seed Is a Book is derived from an article published in *Orion,* March/April 2013, on the sculptures of Basia Irland, text by Kathleen Dean Moore.

What a Find! is derived from Gary Paul Nabhan's book, *Where Our Food Comes From.* This is the exclamation by Vavilov's colleagues upon discovering a sample of wheat with primitive traits in his rucksack, left behind—his last collection from the field, found while on an expedition into a remote region of the Carpathian Mountains, before he was whisked off in a black sedan by Stalin's emissaries.

Seed Freedom is an expression used often by Vandana Shiva, part of her core message. She speaks with urgency, authority, and grace.

From Chapter 4, *Acrobatic Time:* My son Levin, at age 3, looking up to a sky full of starlings on the Penwith Peninsula came up with this line: "Look at that great space of birds!"

Alchemilla, or lady's mantle, was indeed present in our Cornish cottage garden, but the consonantal assonance of the ll's secured the place for the flower in the poem.

ACKNOWLEDGMENTS

I HAVE HAD THE GOOD FORTUNE to cultivate the fine silt loam of Amagansett, New York, for 24 years. I thank Deborah Light daily for her gift of this land to my employer, the Peconic Land Trust, whose mission is to preserve the land and heritage of the eastern end of Long Island. I have farmed beside many dedicated, sparky young people throughout these years, and I commend them for putting up with "my gab and my loitering" (as another Long Island poet put it). Joe O'Grady deserves a bucket of praise for his studious attention to plants and seeds and food, and for his pesky, delightful humor. Throughout the 2013 growing season, the following spirited crew tended the fields (with "tenderness and gristle," as the poet Lorine Niedecker said), while I tended words: Layton Guenther, Kate Rowe, Honna Riccio, James Walton, Matt Dell, Irene Berkowitz, Gregg Kessler, and Ella Fleming: Thank you.

It was a privilege to serve as mentor to Amanda Merrow and Katie Baldwin of Amber Waves Farm, and now to continue as companions in the field. I treasure my friendship with

Liz Moran, Steve Eaton, and their sprightly daughter, Harbor. To Anne Jones Levine and her lovely children, Willa and Ezra: *May the nourishment of the earth be yours.*

I thank all of my colleagues at the Trust and especially our founder, John Halsey, and Pam Greene, who have given me the space every writer yearns for. Rebecca Chapman has offered continuous support to me and to the other organizations I serve, and I admire her subtle, playful way of dealing with the inevitable complexities of keeping a not-for-profit afloat.

I am grateful to my dear friends who help to guide Quail Hill Farm with imagination and grace: Hilary Leff, Jane Weissman, Linda Lacchia, Kevin Coffey, Gordian Raake, Jerry Pluenneke, Leigh Merinoff, Hope Millholland, Arthur Kaliski, and Julie Resnick. Nick Stephen's gentle banter and his artistry with a hammer are woven into the structures of our community farm. I thank Nancy Goell for our creative conversation and for her gift of listening. It is difficult to accept that Eileen Roaman has left us, but her generosity to young farmers, to me, and to our community lives on.

Several chapters and the proposal for this book were composed at the home of Alan and Edith Seligson on the island of Vinylhaven, Maine: Thank you. Thank you also for lodging and more, along the way, to Krysia Osostowicz and Simon Rowe, Nicola Winter and Lewis Love (two couples who reside in London), and to Kirsten Falke-Boyd, in Berkeley. My lifelong camaraderie with Peter Perry, of Paul, on the Penwith Peninsula, keeps Cornwall close.

My colleagues at NOFA-NY have always inspired my

explorations—in the field and on the page. Deep appreciation to Elizabeth Henderson, Kate Mendenhall, Liana Hoodes, Mary-Rose Livingston, Mark Dunau, Jamie Edelstein, Karen Livingston, Anu Rangarajan, Laura O'Donohue, Robert Hadad, Karen Meara, and Nichelle Wade.

For the creative time spent together on the Board of Sylvester Manor Educational Farm, thank you to Bennett Konesni (and Edith Gawler), Eben and Susan Ostby, Sara Gordon, Edie Landek, Don Shillingburg, Al Kilb, Sam Seymour, David Kamp, Jennifer Ruys, Thomas Carrier, Kathleen Minder, and Stephen Eisenstadt.

For many years I was lucky enough to be a part of the founding board of the Center for Whole Communities in Vermont, and I treasure the friendships formed there. I am grateful to Gil Livingston, Lauret Savoy, Tom Wessels, Danyelle O'Hara, Torri Estrada, Julian Agyeman, and Diana Wright. In a difficult time, two Vermont friends, part of the CWC community, reminded me of the friendship of the seasons and the joy of man's desiring—John Elder and Peter Forbes. For the speedy, soulful journey down the Yampa (the water flow was at an all-time high), thanks forever to Peter Forbes and Helen Whybrow.

Despite my criticism of certain aspects of the land grant system, I value my collaborations and camaraderie with our local Cornell Cooperative Extension, under the leadership of Dale Moyer, and those in NRCS I have had the pleasure to work with.

Three east-end chefs are so close to our mission, our

plants, and our fields, I honor them as family: Joe Realmuto of Nick and Toni's, Colin Ambrose of Estia's Little Kitchen, and Bryan Futerman of Foody's. Leslie McEachern of Angelica Kitchen (12th and 2nd, New York City) has freely given friendship and wise counsel from the beginning of my tenure as a farmer.

This book owes its existence to the inspiration and thoughtful guidance of my agent, Paul Bresnick. I am grateful for his vision, his patience, and for his belief in my work. My former editor at Viking, Paul Slovak, was quick to suggest another book, way back when, and I thank him for his confidence. My present editor at Rodale, Alex Postman, has improved the original concept beyond measure. I hear her now, urging me to press into the language and to burnish it, to be sure I have chosen *le mot juste*. And thank you to the rest of the team at Rodale, who have been so kind and supportive throughout: Kara Plikaitis for her quite beautiful design work, Ruoxi Chen for keeping things in motion, and our senior project editor, Hope Clarke, for pulling it all together.

I am grateful to Maria Bowling for her gift of healing and for the beautiful image that is printed on the cover of this book. It is perhaps enough that we share the same birthday—but we also share an equal respect and reverence (as with Megan) for the beautiful expressions of the natural world that surrounds us.

The subject of seeds is so vast, I have felt humbled from the beginning; I only hope that my field husbandry and the lyric arc of my prose may harmonize with the songlines of seeds. I derive some Celtic comfort from an observation made by my painter/weaver/ jeweler friend, Breon O'Casey, who said of his own (exquisite) work: "At least I have added something to the heap."

My sister, Jane, is the kind of reader every writer desires. Unfortunately, it takes me about 5 years to write a book, while she is committed to reading it in less than 5 days. I promise to do what I can to improve the ratio.

Connie Fox, a painter, and Bill King, a sculptor, my in-laws, and the real thing when it comes to making art, have been an inspiration throughout my adult life. Let's sing it out: "Ars Longa!"

My family has been very much part of the making of *Seedtime* from the beginning. My son Levin kindly gifted me his laptop; my daughter, Rowenna, with grace, offered tech support and heart support; my son Liam, after completing a series of beautiful drawings, asked hopefully, "Can I do more?" Megan helped me to organize thoughts, files, and footnotes; herein my words are the warp, her poetry and music is the weft. Searching for renewal, I found it very close at hand, in the heart of my family. Their names are magical to me, like the miraculous life force within seeds: Thank you!

INDEX